WELDING SECRETS

Second Edition

A welding guide for the self taught welder,
as well as the more experienced welder.

by

HAL WILSON

Published by

FLYCO MACHINE COMPANY
3712 Ringgold Road #155
Chattanooga, TN 37412-1638

Web Site:
http://www.entermall.com/weldingsecrets

Printed in the United States of America

Table of Contents

Foreword

I have seen a lot of welds in my life. I've seen them on new products and on repair jobs. The variety of welds I've seen lately is enough to convince me that some people **still** don't know where **not** to weld.

In this book I've tried to help those of you who will go on welding a long time after I am gone. I've tried to illustrate as clearly as I know how — in words and in pictures — some of the places that should **not** be welded.

I've also included some tips that will help you do a better, safer job in places that should be welded.

I hope this book will help you be a better welder. The information in it — gathered the hard way over 49 years — has certainly helped me.

Hal Wilson

How To Remove A Bad Bearing Race Or Cup From A Hole

If the race or cup is bad, it will be discarded as scrap. Take your arc welder and carefully run a bead around the center of the race, keeping it in the center to prevent drifting off to the other surface. After cooling, it will have loosened enough to fall out.

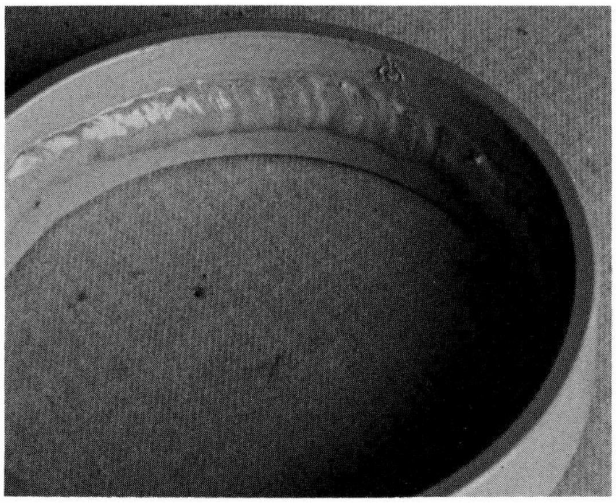

Fig. 1

The above race was 4 7/16 inches in diameter before it was welded. After it cooled it was 7 thousandths of an inch smaller in diameter.

Replace New Race

An easy way to replace the bad race with a new one is to place the new race in the freezing compartment of a refrigerator and freeze it before inserting it. Handle it with gloves.

Occasionally, a race will turn in the hole and wear the hole out too big.

If an internal knurling tool is unavailable, take a center punch and a hammer and make center punch marks all the way around the entire surface that is loose. The amount of wear or looseness determines how hard you must hit the center punch. When a piece of mild steel is hit hard enough to make an indentation, the surrounding metal swells outwardly.

How To Build Up A Worn Shaft

First, turn the worn part off in a lathe. Then, preheat it with a torch to about 250°. Have the shaft in a set of rollers so you can turn it from side to side. Weld a bead on the top side. Turn the shaft a half-turn so the new bead is on the bottom. Then weld a bead on the new top side exactly opposite the first bead. Turn the top of the shaft slightly toward you to make the side of your second bead level. (See Figure 2)

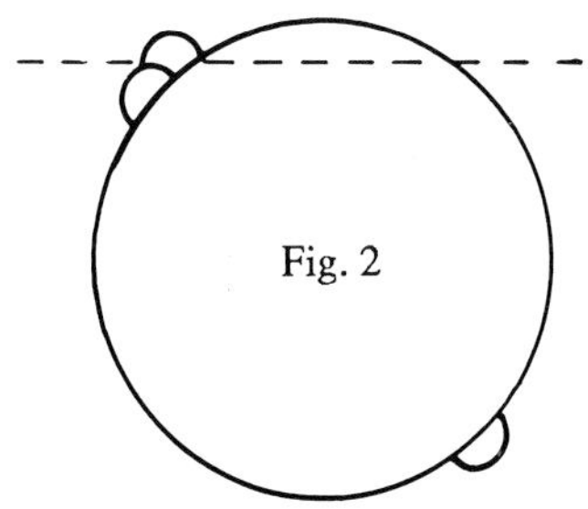

Then run another bead overlapping to the center of the second bead. Turn the shaft back over to the first bead and run another bead overlapping the center of the first bead.

You don't need to knock the flux off. After the second and third beads, turn the shaft after each bead to the other side. On the last two beads, weave the rod from side to side to overlap the first and last beads on that side.

Fig. 3
Shaft turned down ready to be built up.

Fig. 4 shows a built-up shaft. It is hard to believe how much tension is on these welds. They are actually squeezing so hard that the shaft adjoining the welds is compressed to a smaller diameter. I never gave it a thought until I had a shaft break off at the end of the welds. Therefore, you should always stress relieve the welds. As soon as you finish welding, take your acetylene torch and heat the welds to a red heat. Turn the shaft and heat evenly. Hold the heat on it and give it time to get red to the center of the shaft.

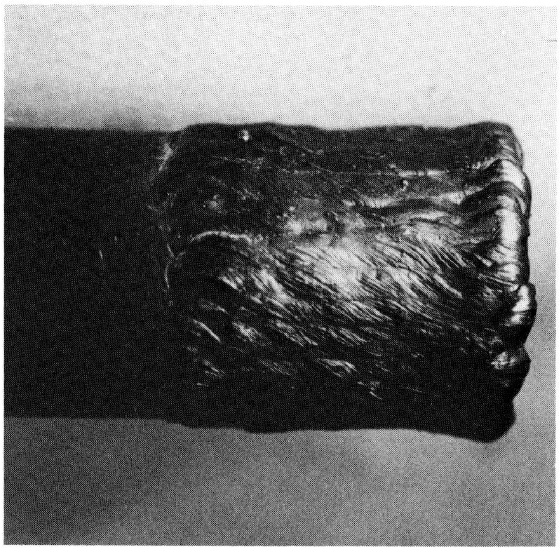

Fig. 4

Fig. 5

To prove a point, I turned the welds down on a lathe, to the original size, freely slipped the bearing race on to the other end of the shaft, stress relieved the end that was welded, and let it cool. When it was cold, it was 3 thousandths larger in diameter than it was before I stress relieved it. The bearing race would not come off. I pressed it part of the way off. You can see where I stopped. Then pressed it back on.

How To Burn A Nut Off Of A Bolt And Save The Threads On The Bolt

It's not as hard as it sounds.

You can't burn the threads on the bolt unless you get them red hot. Place the bolt and nut in a horizontal position. Using a small cutting tip, hold the tip parallel to the bolt. (See Below) Start on one of the points and burn through the nut, cutting parallel to the bolt until you get near the bottom. Then, tilt the end of the cutting tip downward and move rapidly through the slot with your stream of oxygen, washing the threads of the nut out of the threads of the bolt.

You can use the same procedure to remove a piece of broken off pipe from a fitting (steel pipe only). Concentrate the heat on the broken off pipe.

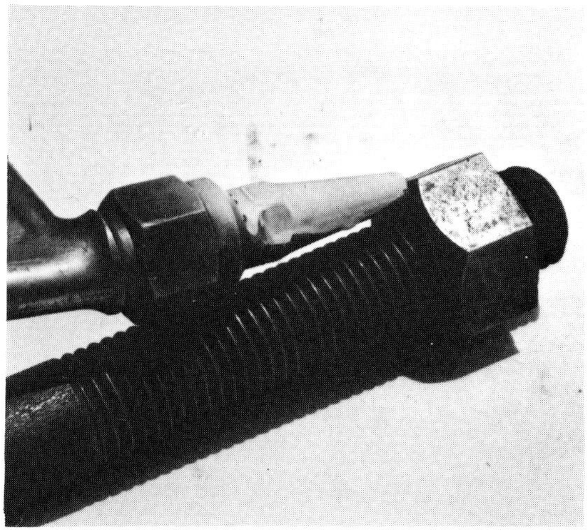

Fig. 6

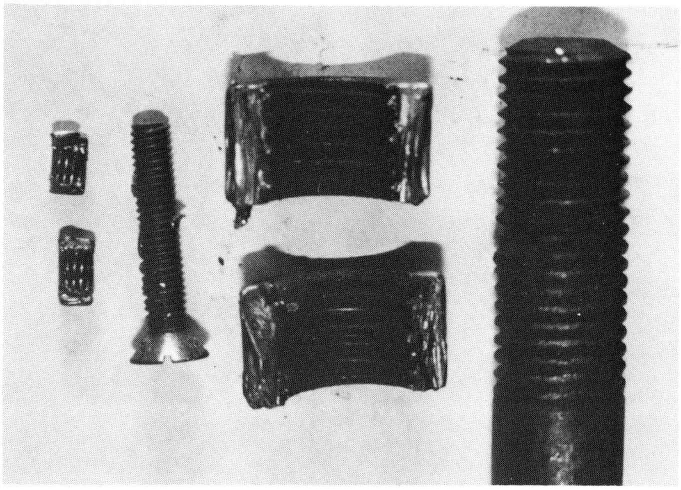

Fig. 7

How To Expand A Nut

To make a nut that is too tight, loose enough to turn with your fingers, do this:

Screw the nut on a bolt. Lay the nut on an anvil and, with a very small ball-peen hammer, peen the flat sides very, very lightly with the ball end. Be extremely careful not to hit it hard enough to make a dent in the flat side of the nut. You can hit it fast, but not hard. Raise the hammer not more than two inches above the nut. If necessary, you may peen all sides that way.

How To Shrink A Nut

When the nut is too loose, do this:

With the nut on the bolt, place the nut in a vise flat side up. Grip the two points of the nut in the jaws of the vise very lightly. Then heat the flat side on top red hot. When it turns black, chill it in water and check it for fit. If necessary, you can heat all six sides this way, one at a time.

Camouflage

The nut below appears to be welded onto the bolt. As you can see, it is not. The nut was welded onto another bolt, cut off, and then drilled to the original "tap drill" size. Then the tap was screwed thru from the back side to tap the weld. This is for larger bolts, the smaller bolts twist off too easily. You can also drill a hole through the nut and bolt, drive a rolled pin through it, putty the ends and paint over it. Occasionally, this procedure may deter a thief.

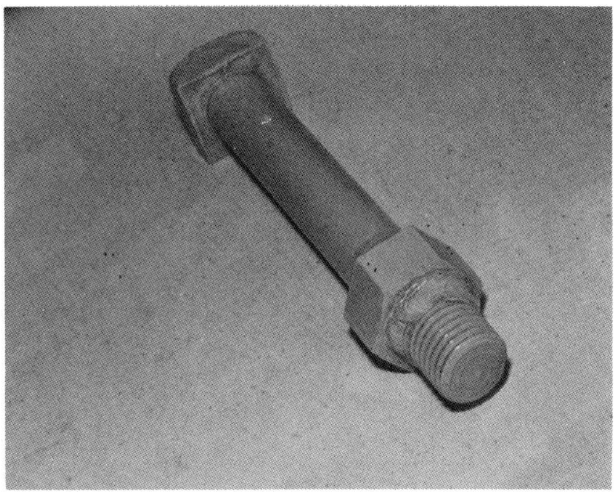

Fig. 8

Fig. 9

How To Remove A Broken-Off Bolt

There are several ways to remove a broken-off bolt. "Ridgid" has a very good screw extractor set #10.

Another way is to grind and break off the bent end of an Allen wrench, then grind across the end until it is square. The points should be sharp. Drill a hole in the broken-off bolt the size of the diameter of the flats on the Allen wrench. Then drive the Allen wrench into the hole. The points will cut the metal out and make a hexagonal hole.

These two methods work well — but if the bolt is steel, you can't beat welding a nut on it (See "How To Remove A Broken-Off Tap," page 13).

For example, our shop was once faced with 16 flat head screws, 1/4 by 3/8 — all stuck. The head had a hole for an Allen wrench. We bent an Allen wrench and stripped the hole in another screw. Then, we tried penetrating oil. We hit them with a punch.

Nothing worked — and only three threads holding. Then we welded a nut on it, using the vise grips as shown in "How To Weld A Screen Wire To A Steel Frame." See page 8) It's the only thing that worked.

The reason this method works so well is that when the bolt is welded, the heat radiates through it. Filling the hole in the nut adds still more heat. When the bolt expands and meets resistance from the hole, something has to give. Either the bolt is compressed or the hole is stretched. Either way, the bolt comes out when it cools.

Shown is one of the 1/4 by 3/8 inch flat head screws with a nut welded to it.

Fig. 10

Case Hardening

Caution: When cutting with a cutting torch, be sure that you do not run out of oxygen. If your oxygen pressure gets too low, a feather will appear on your flame, showing an excess of acetylene. The melted metal on each side of your cut will absorb carbon from the excess acetylene. As you know, when cutting steel with a cutting torch, a very small area gets red hot. The surrounding cold steel absorbs the heat so fast, it is almost like dunking it in water. The result is, the steel that was melted, and absorbed the carbon is case hardened.

I remember 20 years ago, I was cutting some flanges for hydraulic cylinders out of 1/2'' plate with a circle burner. My teen-aged son said, ''Let me do that dad.'' I handed him the torch. Later he came to me and said there was something wrong. He was running out of oxygen. We changed oxygen tanks. When I started to machine the 3 pieces that were case hardened, instead of the lathe tool cutting the flange, the hard place cut a groove in the lathe tool. I heated them red hot and covered them with a bucket to retard the cooling. When they were cold, they were soft enough to machine.

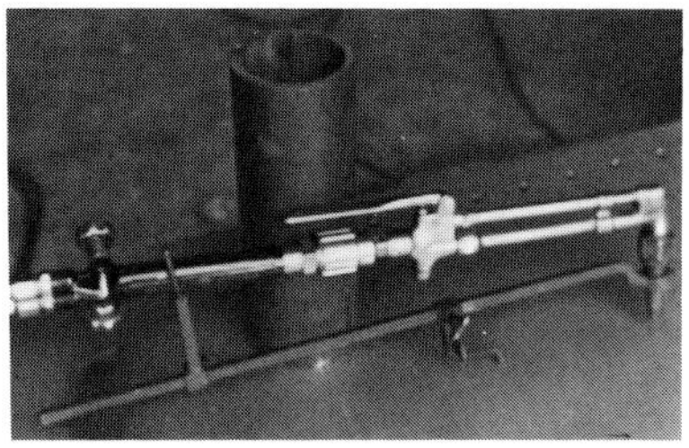

Fig. 11

How To Square A Frame By Welding Or Peening

Every weld in a frame pulls the frame toward being square or out of square. The direction of travel causes it. Every weld pulls the metal closer together at the point where you stop welding than where you started. The first welds on the four corners of a frame should be made in the same direction — from inside to outside or vice versa. After you weld the corners, check the frame for square by measuring diagonally across the corners.

If two corners are farther apart, weld a bead on the other side of the frame from the inside to the outside on both corners. If the frame does not square with that, weld another bead on the other two corners, traveling from outside to inside. In other words, **open** an angle by traveling from inside to outside. **Close** an angle by traveling outside to inside.

You can also open an angle by peening on the inside edge. Close it by peening on the outside edge.

How To Flatten A Warped Steel Frame

Flattening a warped steel frame is simpler than it sounds.

When an angle iron frame is warped — if two diagonal corners are raised when the frame is placed on a flat surface, causing the frame to rock — here's how to solve the problem:

Lay the frame on a flat surface with one leg of the angles pointing down and the other ones pointing toward the inside of the frame. (See Figure 12)

Place an adjustable wrench on the end piece of angle on the leg pointing to the inside of the frame. Place it about one inch from the corner that is higher.

At this point, you have to do just the opposite of what it looks like you should do. You push **down**, not up, on the wrench handle until you can feel it give. (See Figure 13)

Then, do the same thing on the opposite high corner. If the frame still refuses to lie flat, place the tool on the angles on each side as shown in Fig. 15 and 16, and push down. The frame is turned over in Fig. 17. If necessary you may pull inward on the high corners and outward on the low corners on the legs that point upward.

Some angle irons are too large to be bent with an adjustable wrench. For these, cut a slot in a piece of heavy steel and weld a long handle on it. Make the handle strong enough to bend the angle iron leg. This procedure works on many different shapes.

When you bend a piece of cold mild steel, it always springs back a little. If you bend it more, it doesn't spring back any more than it did the first time.

When you bend the leg on an angle to flatten a frame, the leg springs back a little. That puts a torsion stress on the angle iron causing it to push down on the high corner. It isn't necessary to bend the leg too much. You have 16 points on an angle iron frame, (four on each corner) to flatten it with.

The iron-worker who taught me how to do this also told me a story:

A new man came to work at the shop where my friend worked one day. The new man built a gate out of very heavy, closely woven wire. When he finished it, it was warped. He worked all morning, hammering and twisting. He even blocked up the low corners and pressed down on the high corners. It flopped — and warped the other way. What had been the high corners were now the low corners.

My friend watched out of the corner of his eye until the new man took off for lunch.

"Then I went over there and straightened it out with a pair of pliers," said my friend.

When the new man came back from lunch, he checked all four corners. They were all flat on the floor. He turned the gate over and checked it. He stared at it, scratched his head, looked all around the shop and demanded, "Who straightened this for me?"

He never did find out.

My friend's secret? He put stress on the wires near the frame with his pliers. Just the opposite from the way it looks.

Fig. 12

Fig. 13

Fig. 14

Notice the clock. I placed the tool, pushed down on the handle, left the tool in place and then made the picture, on each one of these. I straightened one without making pictures in less than one minute.

Fig. 15

Fig. 16

Fig. 17

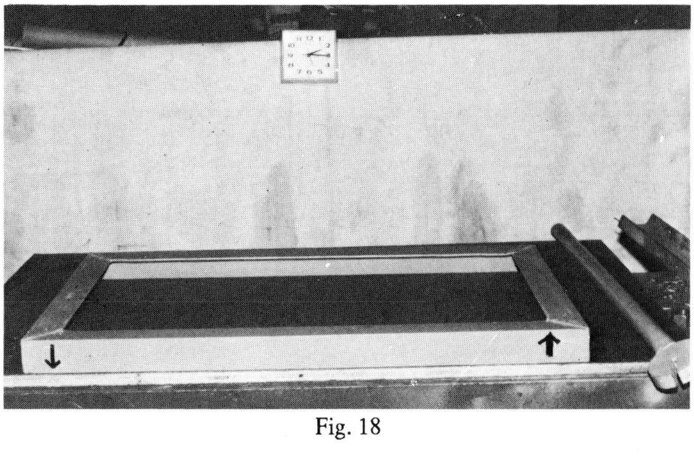

Fig. 18

How To Weld A Screen Wire To A Steel Frame

If you're having trouble welding a screen wire or light expanded metal to a frame, try using washers.

Place a flat washer on the wire where you want to weld it to the frame. Clamp it down with a pair of vise-grips. (See Figure 19) The big, partly cut-off washer is added on, lightly tacked.

Weld through the hole in the washer and into the frame. Fill the holes in the washers with weld metal. (See Figure 20) If you want a really neat job, cut a strip of 1/8 x 3/4 inch to the correct length and drill 5/16 inch holes in the center of it. Clamp it in place and weld through the holes. (See Figures 21 and 22)

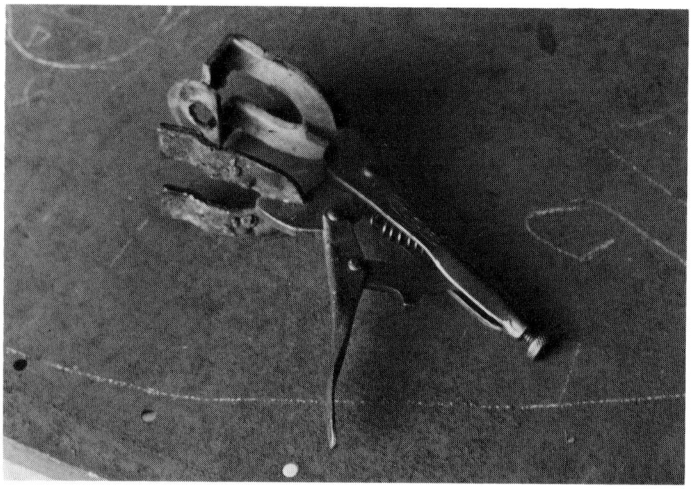

Fig. 19

Fig. 21

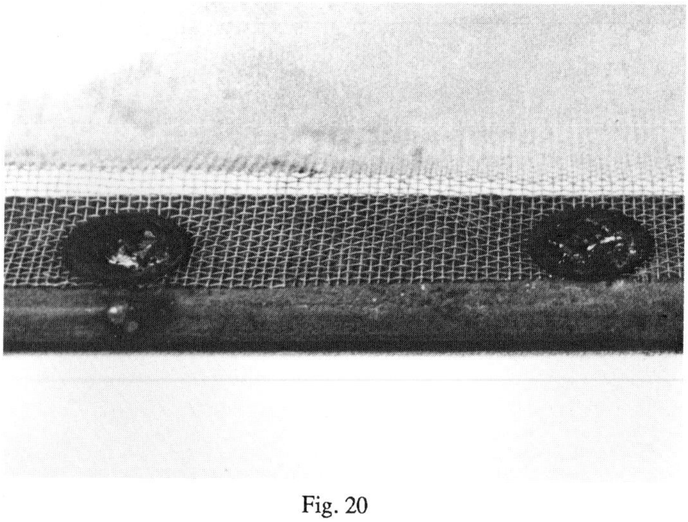

Fig. 20

Fig. 22

How To Make A Smooth Cut With A Cutting Torch

For a machine-like cut, use a straight edge. Take a 1/2 by 2-inch piece about 12 to 14 inches long and a 1/8 by 3-inch piece the same length. Weld or tack the 1/8-inch piece on top of the 1/2-inch piece, letting the smaller one overhang a half-inch on each side. File the 1/8-inch edges smooth. Place the file flat on the 1/8-inch edge lengthwise to keep it straight. Keep the edge filed and sanded to keep it smooth.

To cut, place straight edge far enough away from the cutting line to make the center of the cutting tip directly over the cutting line when the tip is rested vertically and lightly against the straight edge. Drag the torch toward you. You will not be able to see the flow of metal. If all the sparks and fire are landing on your feet, you're traveling at the right speed.

If the slag on the bottom of the cut is too hard to knock off, you are using too much heat. Cut down on the flame, use a smaller tip, or travel a little faster.

Incidentally, the acetylene gas is not what does the cutting. When you pull the trigger on red hot steel, the oxygen attacks the carbon in the steel. The carbon burns and supplies most of the heat. You can prove it by burning the end off a piece of 1/4 by 2-inches. Quickly start another cut on the end. Have your hand on the acetylene valve. Get the cut started and turn off the valve. You can finish the cut without a flame.

That's why you can't cut other metals or even stainless steel with a cutting torch.

How To Do Overhead Welding

Never try an overhead weld with a cold machine. Have the machine hot enough to melt the base metal almost instantly. If your machine isn't hot enough, you'll have to stay in one place too long to melt the base metal and your weld will start bulging downward — it may even drop off. Lay in a little rod and move it away for an instant, before it starts to bulge downward. This lets the crater chill. Then go back, but not as far as you were before.

You can also run light stringer beads. Hold a short arc and move along fast enough to keep it from bulging, but slow enough to fuse.

How To Weld Thick To Thin

To weld a thick piece of steel to a thin one without burning a hole in the thin one, try concentrating the arc entirely on the thick piece, moving slowly toward the thin piece with the crater edge until it fuses to the thin piece. Then back away before it burns a hole.

How To Make A Smooth Bead

Every welder is judged by the looks of his work.

To run a smooth bead, position is very important. Get comfortable. Use a rest for your left hand. With the electrode holder in your right hand, rest your right hand on your left hand so that you can burn a complete rod in a smooth, steady movement. Practice the movement without the rod.

Vertical Weld

If a vertical weld bulges in the center or drops off, you are probably coming back down too low. When you raise your rod to give the crater a chance to chill (freeze or solidify), you are coming back down too far. Try this method:

Bring the end of your rod back down and across the top edge of the crater that just chilled. You can also run a vertical bead downward. Use a fairly hot machine, tilt your rod upward, and run a light bead with a close (or short) arc. Travel fast enough to keep the flux and melted weld metal from trying to get ahead of or below the end of your welding rod, but slowly enough to get fusion.

How To Flux Lead

If you have a pot of melted lead and it has a lot of skimmings on top of the lead, you can skim it back and it will be a different color. The color changes back rapidly to the original color. This is because it oxidizes when it comes in contact with the oxygen in the air. Don't skim the skimmings off and throw them away. Sprinkle a little pine rosin on the top of the skimmings and stir it up. The pine rosin is a flux for the lead. It breaks the oxide and it all melts. This also works on babbitt and solder - you have heard of rosin core solder.

Bulge

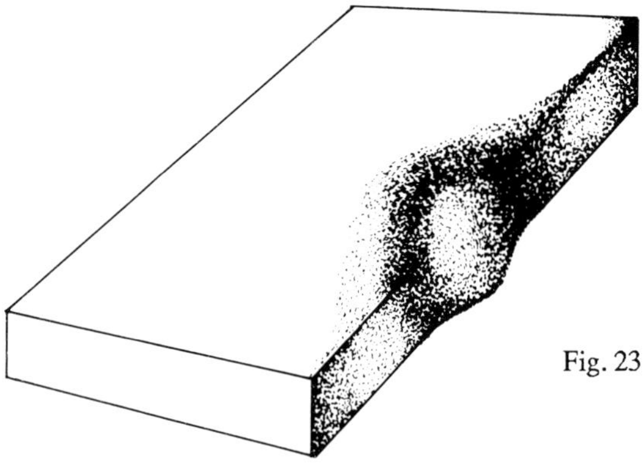

Fig. 23

The bulge in this sketch is highly exaggerated.

If you can imagine that this really happens when a piece of steel is heated red hot in one spot, it will help you figure out a lot of problems that arise in expansion and contraction.

When you start heating one spot, the steel begins to expand on that side. The center of the strip will be under tension and the other side will be under compression. But when the spot gets red hot it has very little strength. The red hot part is pushed together causing it to swell. When it starts to cool and begins to turn black it has more strength. As it contracts, one side of the steel becomes shorter than the other. And the strip bends edgewise.

How To Burn A Weld And Save Both Pieces

If, for instance, you have two quarter-inch pieces lap welded to each other, here's how to handle it:

Light your cutting torch, then cut the oxygen pressure down at the regulator — to the point where you can barely hear it when you pull the trigger.

Position is crucial. Position yourself slightly to the left of the work. Hold the torch with your left hand near the tip. (Have some sort of support handy to rest your left hand on.) Hold the torch with your right hand ready to pull the trigger. Start at the left and travel right. Place your head where you can see the flow of metal. The tip of the torch should be at about a 35-degree angle. Heat the corner red hot and pull the trigger.

When the flow of metal goes through the weld, you will see a black spot at the bottom of the flow. It's okay. Keep on going. If the flow of metal flies back in your face, the oxygen pressure is too high. If you don't see that black spot at the bottom of the flow, you are cutting through the weld and into the other piece of quarter-inch plate. Tilt the tip down more and get below the weld with the flow of metal.

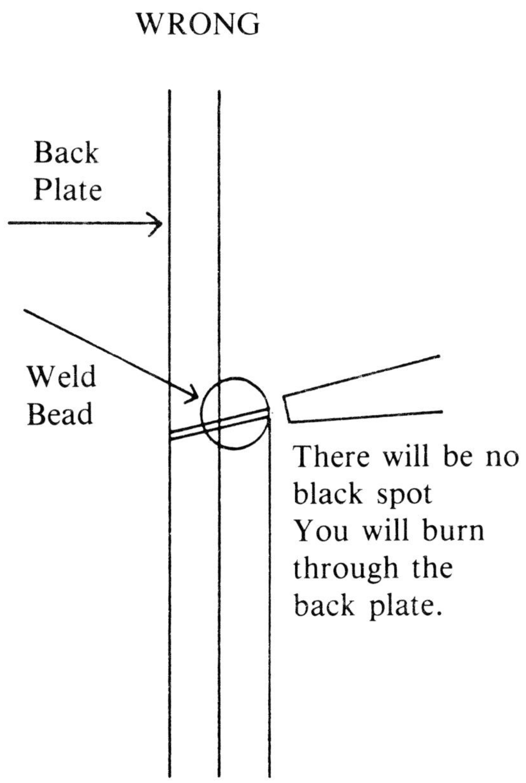

Fig. 24

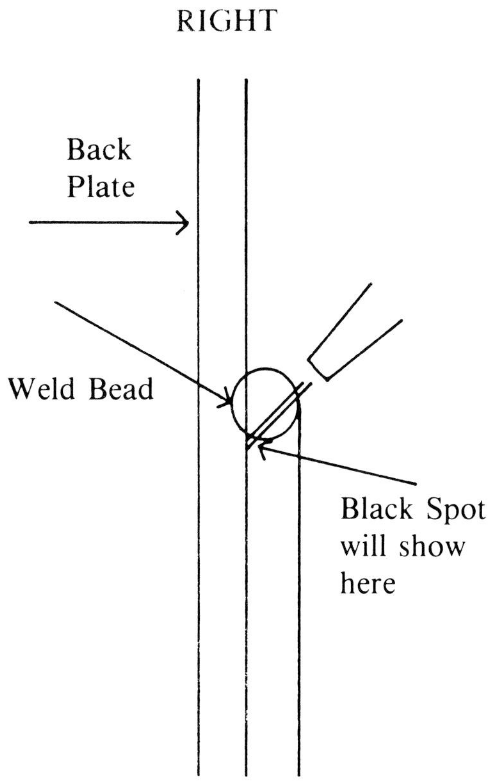

Fig. 25

Big John

Big John had been working at the shop for 16 years when I started my apprenticeship. He was a top notch welder. He welded all of the broken truck frames — and there were a lot of them back then! Big John had his own way of reinforcing them. He didn't put a fishplate on the side of the channel as you might expect. After he welded the crack up, he would weld a flat bar (about 1/4 x 2-inches or thereabouts) on the top or bottom leg of the channel, extending six or eight inches on each side of the weld. Some of the frames started cracking at the top if the load was too far to the rear of the rear axle.

Big John argued that the weld would hold as much weight as any other part of the channel, and if the reinforcement was put on the bottom (or top) leg of the channel it would prevent the crack from starting in the first place. And if the crack never starts, Big John reasoned, it will never break. He told me, "I have welded hundreds of them like that and I have never had one of them come back on me yet." I never doubted him.

Once I saw Big John brazing a whole pile of little cast iron frames. I noticed that he was not preheating them. The frames were about 1/8 inch thick by 4 inches wide and 6 inches long. The sections were about 3/4 inches wide — one on each side, one on each end, and one across the center. They were all broken on the center section. He had previously ground the cracks out. I asked him why he didn't preheat them.

He told me, but only after I twisted his arm. Here's his trick:

Big John held the torch at an angle so that, while he as heating the crack on the center section, the other part of the flame extended over to the end section and heated it. That let the two sections expand together and then shrink together. If I had been doing it, I would have heated all three cross sections.

I learned a lot from Big John.

Another time, I was helping him with a big steel flywheel, about six or eight feet across with a heavy rim and spokes. We blocked it up on bricks to keep it off the dirt floor. Then, we placed several natural gas torches around it and started heating it with soft flames (no air pressure), on the rim only and covered it with some sheet metal. The heat had not been on it for more than five or ten minutes when there was an explosion that shook the building.

The boss came running out of his office to see what catastrophe had hit. By that time, Big John had removed the sheet metal and cut the torches off. One of the spokes had pulled in two, leaving a wide gap (the spokes were oval, three or four inches thick and ten or twelve inches wide — solid steel).

"John," the boss growled, "you heated the rim too hot."

"Look boss," Big John said, "if that crack goes back together when it cools, you can have my job. But if it doesn't, don't open your mouth."

Here's what happened.

The guys who removed the flywheel had trouble getting it off the shaft. They heated the hub of the flywheel red hot in order to remove it. In doing so, the hub tried to expand outwardly, but it couldn't. The rim and spokes were cold. They removed the flywheel and the hub began to shrink. The shrinking of the hub put tons and tons of tension stress on all of the spokes. When just a little heat was added to the rim, which was under compression, it expanded enough to pull the weakest spoke in two. The gap did not close up when it cooled.

Notch Effect

Cut a notch in a shaft, and you know where it will break - in the notch. This is called a "notch effect." There are many others. A shaft turned down in one place to a smaller diameter will break right next to the larger diameter.

A built-up shaft will break at the end of the welds. A poor weld is a notch effect. A shoulder of almost any kind is a notch effect. The end of a reinforcement is a notch effect which can be minimized by tapering the ends.

How To Remove A Broken Tap

Place a nut over the tap and weld the tap and the nut together. Fill the hole in the nut with weld metal. Use a reverse rod with a light flux. Let it cool.

Before you start, the nut needs to be held in place. One way to do this is to use a washer with a hole bigger than the hole in the nut. Weld it to the end of a flat strip of metal. Place the washer over the nut. Block up the other end of the flat strip to make it level and place a weight on the flat strip. (See Below)

With a little skill, a broken off tap can be removed with a cutting torch. Concentrate the heat on the tap only. Hold the tip in a straight line with the tap. Let about half of the flames go down the side of one of the flutes, and the other half on the center part of the tap. When the tap gets red hot, pull the trigger and make a small circle with your tip. You will need enough oxygen pressure to blow the melted metal back out if it is a blind hole.

Fig. 26

Built In Stresses

Every weld contracts or shrinks in all directions. Steel is made up of round crystals and expands when it is hot. When a piece of steel is heated uniformly to a red heat, it is free to expand. But if it is heated in one spot, it is not free to expand. When it tries to expand outwardly the cold metal around it prevents it. The crystals are rearranged to make the metal thicker at that point. Then, when the metal cools, it contracts. This puts the crystals under tension. This will cause a thin sheet of metal to buckle. All welds are under stress unless the stress has been relieved. A light bead on a very heavy piece of steel usually cracks the weld wide open. If the hub of a cast iron wheel is heated red hot, it usually pulls one of the spokes in two.

However steel sometimes has stresses already built up in it. When you weld, you may relieve more stresses than you create. Anytime a piece doesn't move the way you expect it to, blame it on the built-in stresses. When a machinist takes a cut on a straight shaft, the shaft sometimes bends a little due to the stresses being cut off. That is probably why the ball-peen hammer was invented. You can peen the inside of a curve on a shaft to straighten it.

If you repair a broken piece of steel you should reinforce it with another piece of steel, after welding it back together. But **never, never** weld the ends of a reinforcement. If the ends are welded, it puts great stress on the metal adjoining the weld. A little more stress and vibration will cause a crack right next to the weld. Cut the ends of the reinforcement on a 45° angle instead of a 90° angle. Always skip weld the reinforcement — weld an inch and skip an inch. Never weld all the way to the ends of a reinforcement. Leave about a quarter-inch, or the thickness of the reinforcement.

Crystals

When a machine part is subjected to severe stresses over a period of years, sometimes the crystals become elongated. If a part breaks, they say that the steel crystalized. I was told that if you heat a piece of steel to a red heat, the crystals would return to their original shape.

To prove this to myself, I cut some 1/2'' round, mild steel bars. I hammered one of them cold, on the end, to bring it to a point. It didn't taper much until it cracked lengthways. The crystals in the steel elongate when the steel is stretched, but they will only elongate so far and then they separate, and that is a crack. Then I hammered another one, but I stopped before it cracked, then heated it to a red heat and waited until it was cold before hammering again.

I continued this process a number of times. The center of the bar did not stretch much, if any. A hole was left in the center. It seems to be quite evident that the crystals return to their original shape when heated red hot.

How To Burn Through A Large Steel Shaft

You will need the biggest cutting tip you have and all the oxygen pressure your regulator will give you. If your tip is big enough and your oxygen strong enough, you can cut right through. But if the tip is just a little too small for the job, try this:

You must be able to see the flow of metal all the way through the shaft at all times. With the shaft in a horizontal position, place yourself on one side of it. The cutting tip should be on about a 45° angle from vertical. You should be able to see over the torch and watch the flow. Before you light the torch, practice moving it in the path that it will travel. Then light it and heat a starting spot. Don't be in a hurry to start cutting; get it red hot. When you do start cutting, weave right and left and make a gap about three-quarters of an inch wide (for a four-inch shaft, wider for a larger shaft). If the gap is not wide enough and you can't burn all the way through, go back and start over again on one side. Try burning off another half-inch. If that doesn't work, your tip may be too small.

Fig. 27

My First Commercial Weld

In 1936 I was a welder's helper. I had been practicing running a bead for some time. One day someone brought in a broken bumper support from a Model T Ford into the shop. The welder welded it back together and tacked a reinforcement over the weld and looked at me and said: "You can weld this, son. Go ahead."

I welded the top and bottom and both ends. Later, the vice president of the company came through and stopped to talk with the welder. In a few minutes he noticed my weld. He jumped straight up and yelled, "Who did that?"

The welder looked at it and realized what I had done. He said, "It's all my fault, boss. I didn't tell him not to weld the ends. Don't worry about it. I will cut it off and put a longer piece on it."

The boss cooled down. And that was the only time I **ever** welded the ends of a reinforcement.

Cast Iron Welded With Nickel Electrodes

If you have a cast iron ornament that was broken accidentally, you can successfully weld it with a nickel electrode. The cast iron should be preheated. However if a cast iron part is subjected to stress and you weld it with a nickel electrode, it usually breaks right next to the weld. I am not a metallurgist and I don't know why, but I think it has something to do with getting the cast iron too hot. Cast iron can be successfully brazed with a bronze rod and an acetylene torch. Chamfer the break and grind 3/4 inches clean on all sides. The bronze will have to be built up a little bigger to equal the strength of the cast iron. Heat the cast iron to a dull red, heat the rod and dip it into "Cast Iron Brazing Flux" and melt it into the Vee. It has to tin the surface of the cast iron (spread out) if it balls up, it will not stick. Fill the Vee and then spread out over the edges that you ground clean. Dip your rod in the flux whenever it is necessary to keep the bronze tinning and flowing into the pores of the cast iron.

Now this procedure is for a single piece of cast iron that is free to expand and contract without putting a stress on another place. On a wheel, pulley, or gear, they can be brazed if you heat them slowly and uniformly to a dull red heat. Some welders can heat the rim of a wheel at each spoke (that allows the heat to radiate each way on the rim and inward on the spokes). As the rim expands and gets larger in diameter, the spokes expand to keep out stresses.

At some places, they weld cast iron blocks and heads for Diesel engines. They grind the cracks out then place the block in a portable furnace. Heat it slowly to a red heat, open a door on top leaving the heaters on, they reach down in the furnace and acetylene weld it with a cast iron rod. Then they close the door and leave the heat on for awhile. When they cut the heat off they leave the block in the furnace. The block is surrounded by red hot fire bricks. It takes some time for it to cool. Then they machine it. It costs about half as much as a new block.

A foundry made me a cast iron sheave with an extra large hub in order to bolt a brake drum to it. When I picked it up, it had a crack in one of the spokes. I showed it to the foreman. He said the mold should have been stripped. He explained that when it took the hub longer to cool than the rim and spokes, they were supposed to remove the sand off of the hub to make it cool as fast as the metal on the rim and spokes. The rim cooled first and shrank, forcing the spokes to squeeze the hub together, making it thicker. Then when the hub cooled and shrank, the rim would not "give" and the hub pulled one of the spokes in two.

Straighten 6 Inch Strip

If you cut a 6" strip off of a 1/2" plate. 8 feet long, with a cutting torch, it will be curved. The side that you cut contracted and now it is shorter than the other side. It can be straightened by stretching the side that you cut, by peening or by shrinking the other side. Stand on edge and with an acetylene torch heat the edge red hot. As it gets red move along. Have someone come behind you cooling it with water.

Clevis On Channel Iron

Figure 28 shows how **NOT** to weld a clevis on to a stress member.

Fig. 28

Fig. 29

Figure 29 shows a better way to weld it. The two pieces with the holes are welded on to the patch, then the patch is welded on to the channel iron lengthways with the channel. No doubt you are tired of reading "Do not weld the ends."

How To Temper A Chisel

The cutting edge of a chisel should be soft enough to be filed with a file, but no softer. If it is too hard it will break.

You will need a tub of water, a file, a pair of tongs or vice-grips, and a torch.

Slowly heat the chisel from the cutting edge back about three or four inches to a red heat. With the tongs holding the chisel on the back end in a straight line with the chisel, place the tip of the chisel into the water about a half-inch or more. Hold it there until it cools. Then rapidly plunge the entire chisel into the water and back out. Rest the chisel on the edge of the tub and, with the file, rub the end of the chisel. It should be too hard for the file to cut. Keep stroking the tip until the file starts cutting. Then, instantly shove the whole chisel into the water and back out again. The tip should again be too hard for the file to cut. Continue filing and dunking until it is soft enough to file when you take it out of the water. Then place the chisel in the water and move it around until it is cold.

When I was serving my apprenticeship, a contractor would send in an armful of drills every few weeks to be drawn out and tempered. The drills were about one and a quarter inches in diameter and about 18 inches long. They were used in an air hammer to break rock and concrete.

A friend had told me how to temper a chisel. I decided to try it out on these pointed drills. Without permission.

A few weeks later they brought in another batch with special instructions that the same man who tempered the last ones to temper these.

It worked.

How To Make A Circle Lie Flat

An 18-inch circle that has been burned out of a 1/8-inch plate will bulge in the center. It does so because the outside edge that was heated to a red heat is under tension. To flatten it, remove the slag, place the edge of the circle on an anvil, and hammer it with the flat face of the hammer centered on the edge of the circle. Hold the other side of the circle up so the edge will be in flat contact with the anvil.

Turn the circle and hit the edge all the way around, hard enough to stretch the edge, but not hard enough to distort the metal too much. Try leaving a little space between each stroke and, if that isn't enough, go around again, hitting between the last strokes. If the edge is stretched too much it will buckle. The buckle may be corrected by heating red hot on the edge in very small spots. Cool each spot with water before heating another one. Or you can skip from one side to the other, letting someone else cool the spot you just heated.

On the deck of a ship which has many buckles, one man heats spots and another one cools them with air and water.

How To Remove A Stuck Sleeve From A Hole

Screw a tap that is slightly larger than the hole in the sleeve into the sleeve — it needs only to scratch the inside of the sleeve. Drive it out from the other end with a mild steel rod (never hit a hard surface with another hard surface).

Another way to remove a stuck sleeve is to heat the sleeve red hot and let it cool. You can heat a large sleeve red hot in a straight line from one end to the other — it does not have to be red hot all at the same time. Heat red hot in one spot and, as it reaches red hot stage, move forward. The black part behind you has already shrunk.

This works on steel. Don't heat cast iron unless you are sure you know what you're doing.

How To Shrink A Steel Pulley

We once had a steel fabricated pulley that had a tapered hole in the hub to fit a split Q.D. hub. The Q.D. hub had a flange which would pull up against the hub on the pulley without being tight on the shaft.

To fix it we heated the rim of the pulley red hot with a torch and slowly moved toward the center, heating it red hot as we went — heating the hub red hot in one place also. It helped, but it wasn't enough.

When we heated it in four places on quarters it worked fine. There was a quarter-inch of space between the flange and the pulley and the hub was tight on the shaft.

CAUTION: do not try this method on a cast iron pulley.

Reinforce Inside Of Tubing Or Pipe

If you reinforce a broken-off piece of tubing with a smaller piece of tubing or solid bar that fits on the inside of the broken off tubing, do not weld through the broken piece and into the smaller reinforcement on the inside. If you do, it puts it under stress and may cause a break. You can see that if you run a light bead around a solid shaft, the light bead has to shrink, but it will have a hard time compressing all of that cold steel on the inside of the small piece.

Cut slots lengthways on the broken tubing to weld through and into the reinforcement before you put the reinforcement inside. The size of the slots depend on the size of the tubing. There should be enough welds on each side of your weld (that welds the outside piece back together) to equal the strength of the tubing. Stagger the slots. With the reinforcement in place and the broken tubing lined up, weld the broken tubing back together. Try not to weld through to the inside piece of tubing. Let it cool, then weld up the slots.

How To Repair A Crack In A Piece Of Tubing

Cut out piece as shown in Fig. 30 - shaded area.

Then cut out patch to fit hole, out of same size tubing. Fit patch in hole smoothly and tack at each end, A and B. Then weld one side only — from D to B — C to D, then A to C. Let cool, then weld other side — same way.

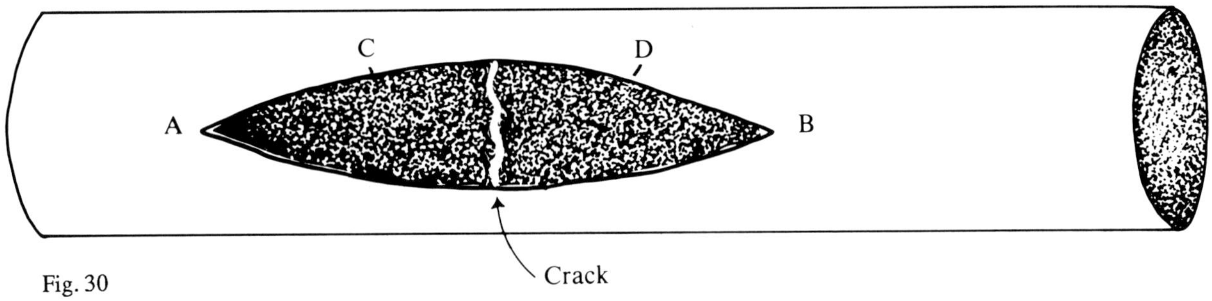

Fig. 30

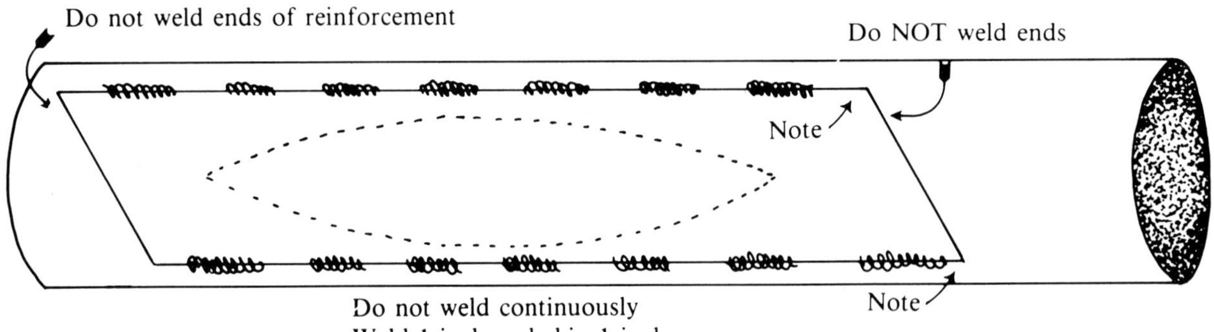

Fig. 31

Grind welds on patch and place reinforcement over patch as shown in Fig. 31. It is best to always grind long ways with a stress member. Cross scratches can sometimes cause a crack. Do not weld all the way to the ends of the reinforcement — leave 1/4 inch or the thickness of the reinforcement.

Weld Inserted Pipe

If you weld around the pipe as shown in Fig. 32, you will create a "notch effect" which usually causes a break adjoining the weld.

Slots may be cut in the larger pipe to weld through as shown in Fig. 33 or holes may be drilled to weld through. Slots or holes should be staggered, with enough weld metal to equal the strength of the pipe.

If you don't have room enough for the slots or holes, you may cut slots on the end of the larger pipe as shown in Fig. 34. The direction of travel should be away from the end of the larger pipe. Start your weld slightly to the inside of the slot.

If the joint must be air-tight, the larger pipe can be cut as shown in Fig. 35. When two beads meet at a point or crotch, extend one bead a little further and try to taper it out to nothing. This prevents stopping the shrinkage abruptly.

Some welders may not agree with me on the "notch effect." You think about it. If every weld shrinks, the last part of your bead will shrink more than the first part. The first part of your bead is already black. Even if the weld didn't shrink, there would still be a shoulder.

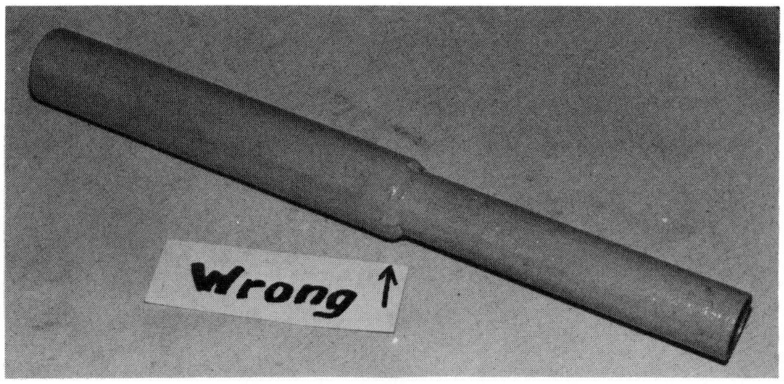

Fig. 32

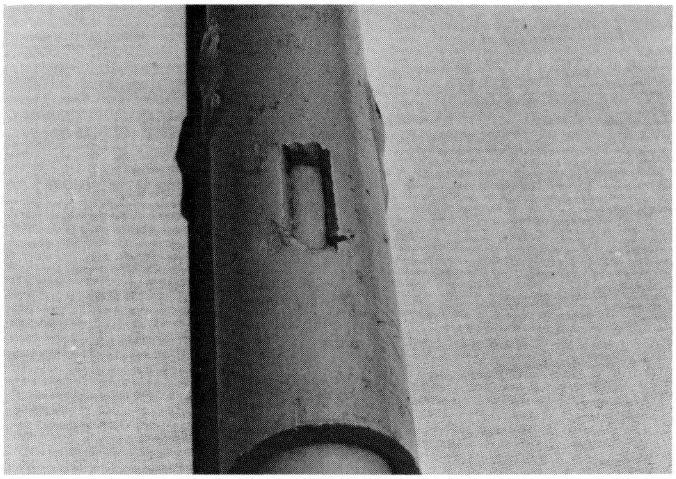

Fig. 33

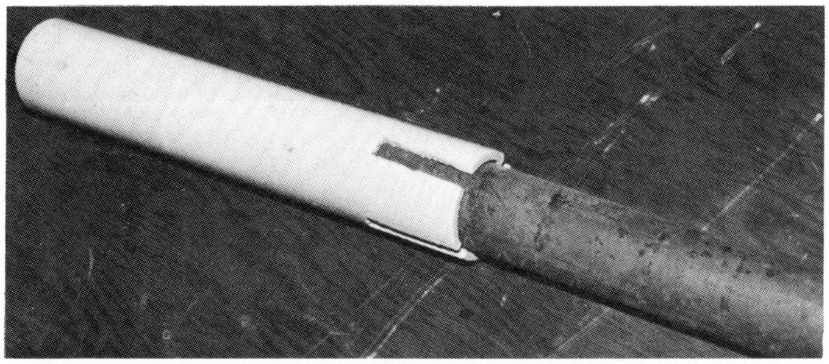

Fig. 34

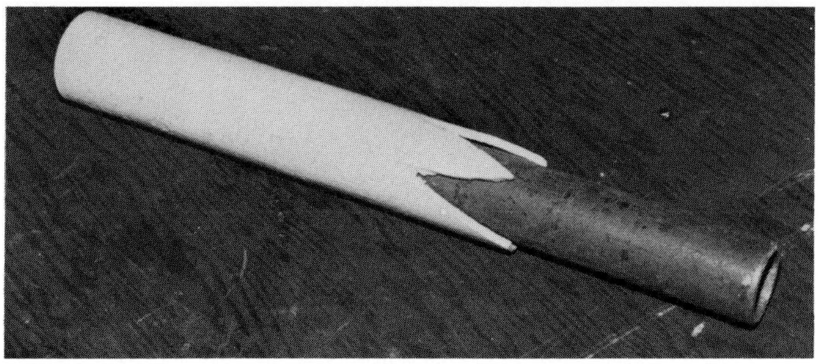

Fig. 35

Another Method For Welding Pipe

It is general practice to bevel two pieces of pipe and butt weld them together. However, if you are using pipe or tubing for structural purposes, and the welds are subjected to tension, torsion, impact or shear stresses, and you have trouble with cracks, you may want to consider the method shown below. The method shown below has no cross welds at all. The welds are back-stepped, and at the ends where the two welds meet, they are reduced to one weld which tapers out to nothing to avoid stopping the shrinkage abruptly. This method is more expensive than a butt weld, but you have a lot more weld metal securing the joint.

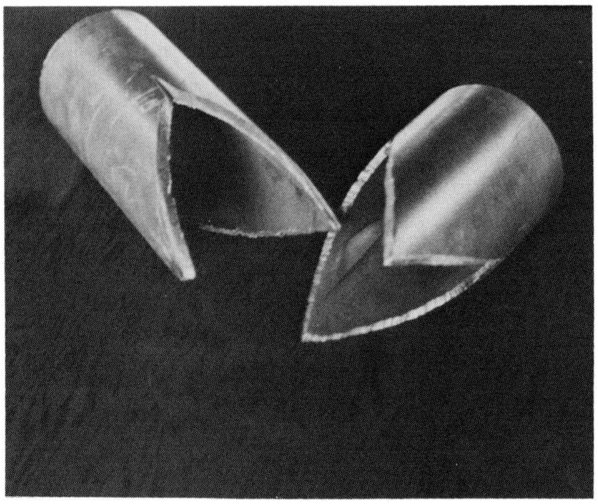

Fig. 36

Make a symmetrical pattern to cut the pipe out with, as shown above.

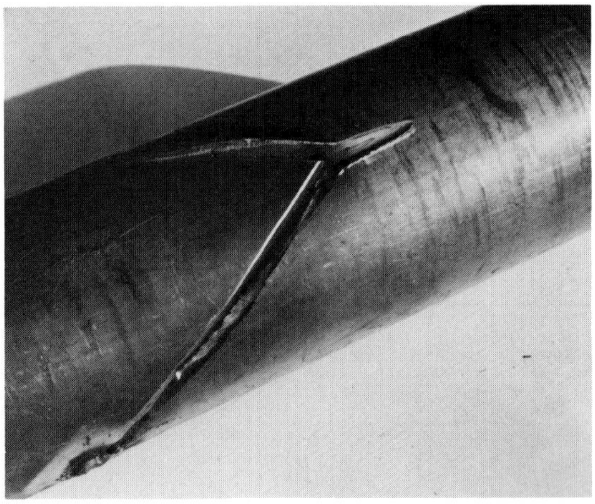

Fig. 37

Bevel edges and grind slot at end.

Another Method For Welding Pipe

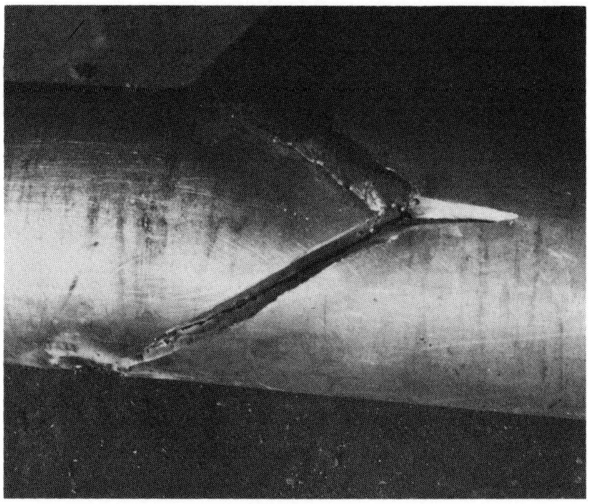

Fig. 38

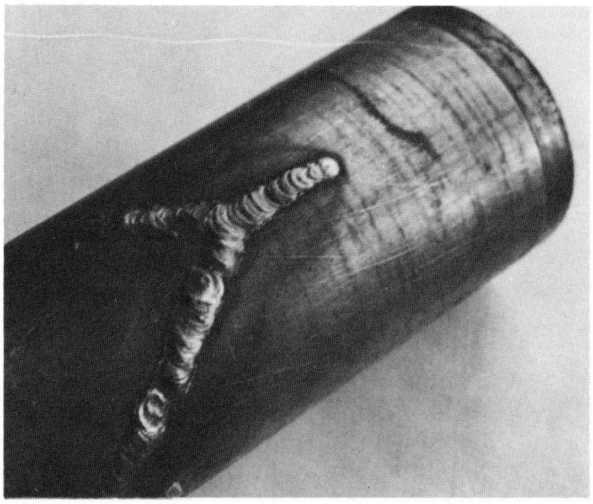

Fig. 39

Weld up two grooves that are exactly opposite of each other and let them cool. The shrinkage will open the other two grooves slightly. Then weld the other grooves. This way, you will have less shrinkage on the joint.

Build Up Inside Of Tubing

The pictures below show that if a tubing is built up on the inside, it will cause the tubing to shrink. This 4 1/4'' O.D. piece shrunk .045'' in diameter.

Fig. 40

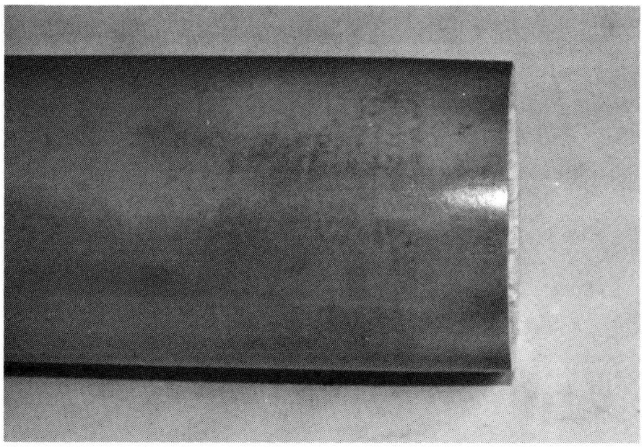

Fig. 41

Test Welds

Figures 42 and 43 are test welds that show how much the tubing shrank from end to end. Figure 42, welded lengthwise, shrank **four** thousandths of an inch in length. Figure 42 probably shrank more crosswise, but that does not cause it to crack.

Fig. 42

Figure 43, welded crosswise, shrank **twelve** thousandths of an inch in length. The cold metal surrounding the welds resisted. The cold metal is now under compression stress and the welds and the adjoining metal that got red hot is under tension stress almost to the yield point.

Think about this and **please** don't weld crosswise on a stress member.

Fig. 43

Flying Coaster

No. 1 shows large gap to prevent someone from welding crossways on the clevis. On the old ride they were welded crossways and they broke off.

No. 2 shows end of patch is not welded.

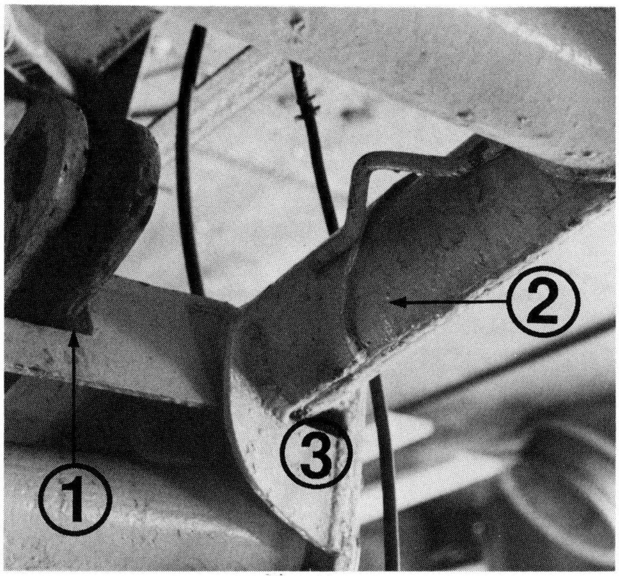

Fig. 44

No. 4 welded crossways, but it is on the inside. I have never seen a failure here.

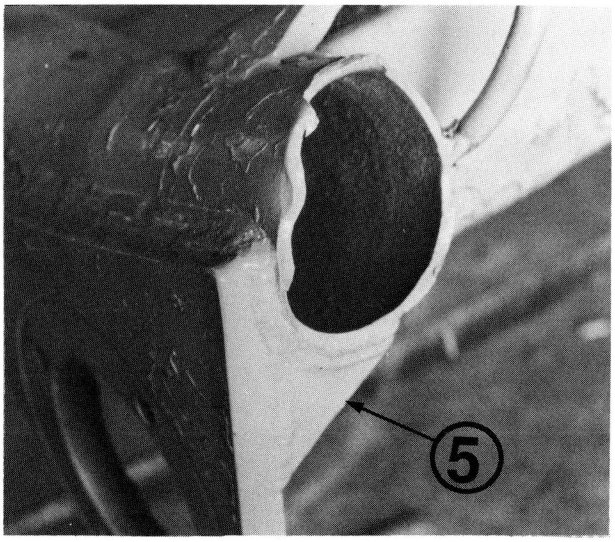

Fig. 45

No. 3 shows that we did not weld a gusset on to the tubing as shown in No. 5.

Fig.46

No. 5 weld on gusset and tubing caused the break on both pieces of tubing.

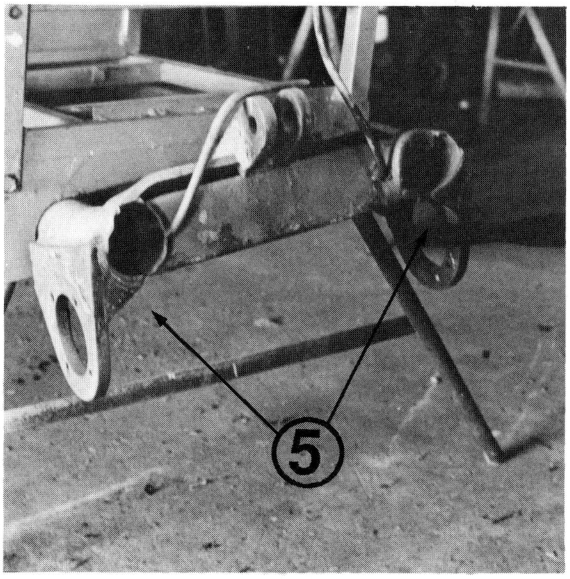

Fig. 47

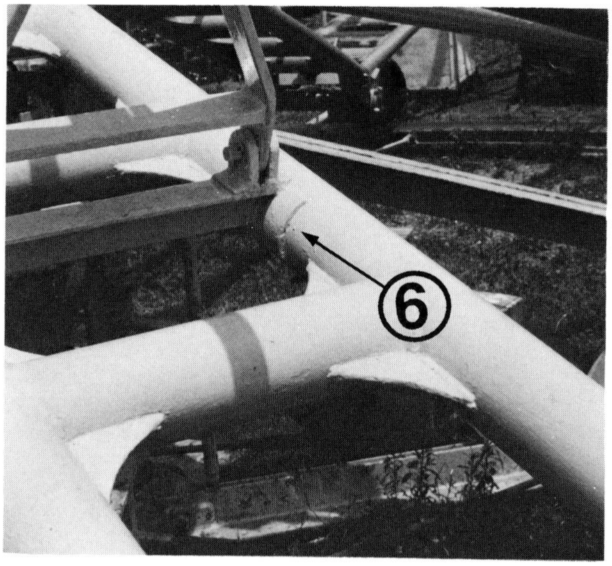

Fig. 48

No. 6 shows T bar welded to patch that is welded lengthways on tubing not welded on ends. On old ride, T bar was welded to tubing and it cracked many times.

Fig. 49

No. 8 shows rod end is tapered on the end, not welded across the end and the welds are tapered. Even a square shoulder here could cause a break.

Fig. 50

No. 7 shows short pieces of angle irons welded lengthways with channel iron.

Roller Coaster

Almost every time I saw a certain type of portable steel roller coaster there was a welder on top of it welding something. With a little investigation, I found out why.

The track pieces were angle irons. The cross ties were channel irons. They were fastened together with very short pieces of angle irons. One leg of the angle iron welded flat on top of the channel iron cross tie and the other welded to the side of the angle iron track. The angle irons were not welded lengthways with the angle iron track, but **crosswise**. (See Figure 52) This causes a notch effect. The angle iron track always cracked right next to the weld. They'd weld it. And it would crack again, right next to the new weld. I actually saw one place with 20 or more welds, side by side, overlapping each other.

Fig. 51

Fig. 52

Cracked Center

This old center for an amusement ride shown below has been abandoned since 1977.

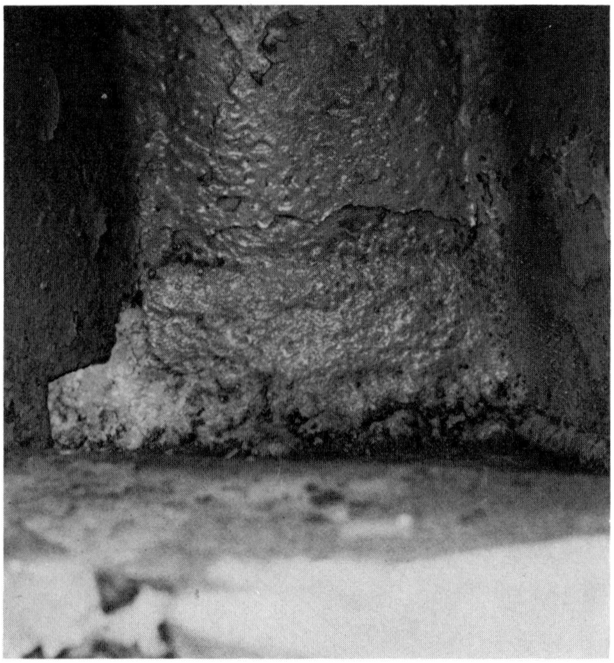

Fig. 53

I remember back in the 1960's, my foreman called me from 600 miles away. He told me the 12'' tubing on the center was cracked. I knew that, with all of the gussets and stiffeners that were welded to it, there could be only one reason for it to crack and that is the shrinkage of the welds. I told him to go find a welding shop that could weld it and call me and let me talk to the welder. I asked the welder if he had an air-hammer. He said ''I think I can get one.'' I told him to vee out the crack and then run light stringer beads and peen each bead with the air hammer before running another bead over it. He said ''OK.'' When the ride was brought back to the shop a few weeks later, it had a big beautiful bead, but it was cracked wide open. The welder couldn't get an air hammer. I had to weld it myself. I peened each bead and also peened the top and bottom beads that were also shrinking. It didn't crack again.

Peening Test

If you don't know how much peening should be done to relieve the stress, there is a simple test you can make. Stand a plate straight up on another plate one inch from the edge. Tack it on and square it to a 90 degree. Weld a bead 2 or 3 inches long on the 1 inch side. Place your square on it to see how much the weld pulled it. Then use an air hammer with a blunt chisel and peen the weld and the adjoining metal which got red hot until the vertical plate is back to a 90 degree.

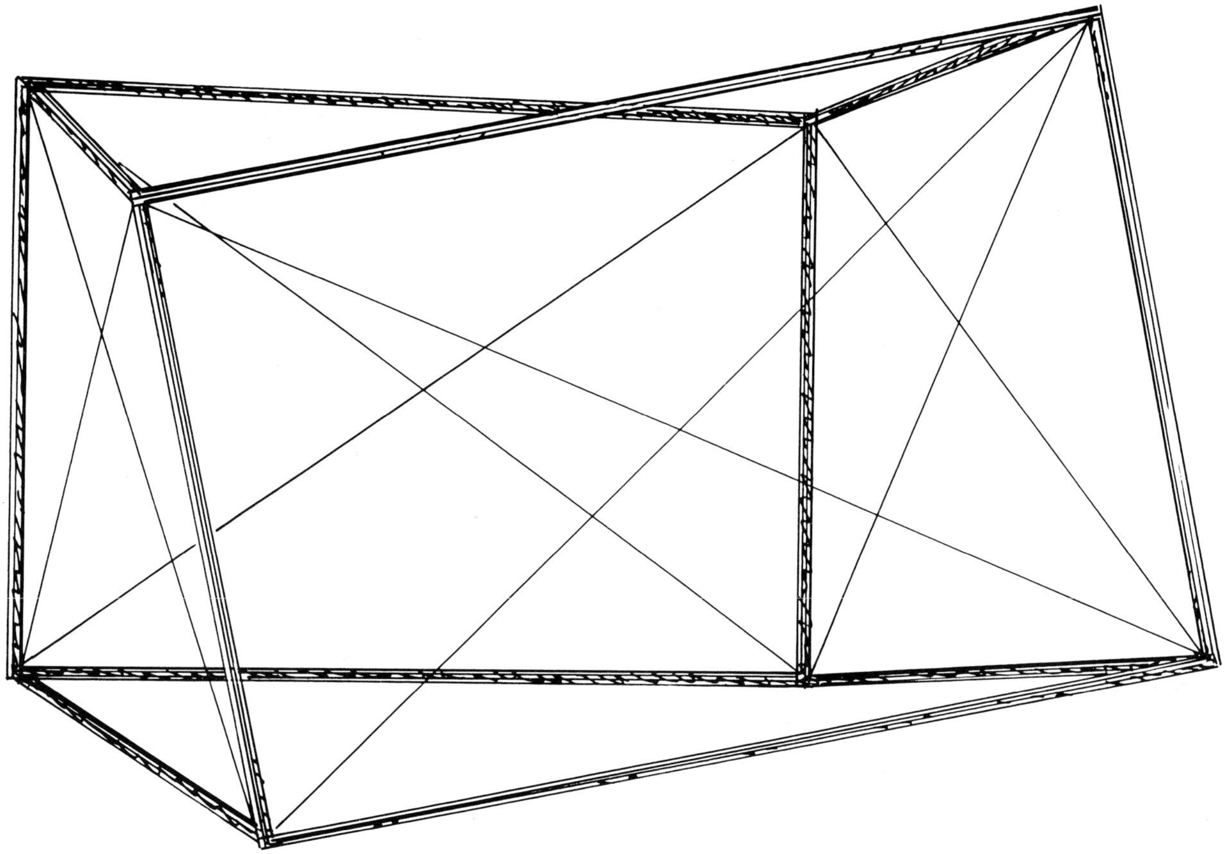

Fig. 54

Cross Brace

A box built of steel, and cross-braced on all four sides will still be flexible. It can be twisted as shown above. To make it rigid, cross-brace the top and bottom also.

For example, twist an open top cardboard box. Two diagonal corners will be farther apart and the other two will be closer together.

Therefore, never cross-brace the top of a truck body. If you do, the cross-brace will tend to lift one wheel off of the ground when it comes to a low place in the road. It could stretch a brace, break something loose or hold the wheel up off of the low place.

Test Welds On Low Carbon Steel

The following subject has been a controversial topic for many years. Whether or not to weld the ends of a reinforcement (or patch).

One writer stated: ''Low carbon steel has such a high ductility that it is *not harmed* by being stressed above its yield point. Hence, it is *not weakened* by any stresses that are left in it after welding.

The distortion and the stresses that are caused by welding have been overemphasized. Perhaps this is due to our zeal to make our welds stronger than the parent metal rather than to make them just strong enough to do the job. So frequently after a weld fails, we ask, ''Where did it break?'' when the question should be, ''At what load did it fail?''

We admit that low carbon steel is very ductile, but almost everything has a limit. The above quotation, in my opinion, has helped cause many failures. I have seen hundreds of reinforcements welded on the ends, with a crack adjoining the weld. I would guess that only about 10% of the welders do *not* weld the ends of a reinforcement.

You can see that if you heat a piece of ½'' round steel in the center, to a red heat, it will expand in length. If it is placed across the jaws of a vice while heating, the ends of the specimen will be free to move outwardly as it is heated and free to move inwardly as it cools. That is ''free expansion''. After it cools there will be little if any difference in the length of the specimen. Now open the jaws of the vice and place the ends of the specimen between the jaws and tighten them very lightly, to restrict the expansion.

As the specimen is heated, it expands and puts compressive stress on the specimen. The specimen will compress a little. But when the spot that you are heating gets red hot, it has very little strength. It will push together and cause a bulge. It will be larger in diameter at the place where it was red hot. As it cools, it will fall out of the vice. It is shorter now than it was before.

This is ''confined expansion''. There is a great difference between ''free expansion'' and ''confined expansion''. Some tests were made to show what happens. A 3/16 inch plate was sawed to about a 6''x 6'' square. Three overlaping beads, 2 inches long were made in the center of the plate. After they cooled, the welds were ground down flush with the plate, to avoid making a ''shoulder effect'' which will cause a crack. There were 2 inches on each end of the welds that were not welded.

The cold metal on each end of the welds resisted the expansion and caused the red hot place to be pushed together (up-set). Now, the places which were red hot are thicker than they would have been if the expansion had been ''free''. As the welds cool, they shrink from all directions, something is forced to stretch. If a weld is not ground down flush, it is thicker than the adjoining metal, a crack usually starts adjoining the weld. If the weld is ground down flush, it usually cracks in the weld.

The 3/16''x 6''x 6'' specimen was bent double on about a 3/4'' radius. It did *not* crack. See Fig. 55.

Then we sawed a specimen from a 3/8'' plate (twice as thick) same size, same welds, same type electrodes and same setting on the machine. We ground it, bent it to an 81 degree and the welds cracked, as shown in Fig. 56. Of course the metal had to stretch much further on the thicker specimen.

Fig. 55

Fig. 56

Then we sawed a specimen 2 inches wide from the 3/8 inch plate. The welds traveled from edge to edge. There was no cold metal on each end of the welds to resist the expansion. Of course, there would be some stress because the welds were not all red hot at the same time. The specimen was ground down and bent double. It did not crack. See Fig. 57.

The specimens in Fig. 56, Fig. 57 and Fig. 58 were all sawed from the same 3/8 inch plate. In Fig. 58 one was sawed across the grain and the other one was

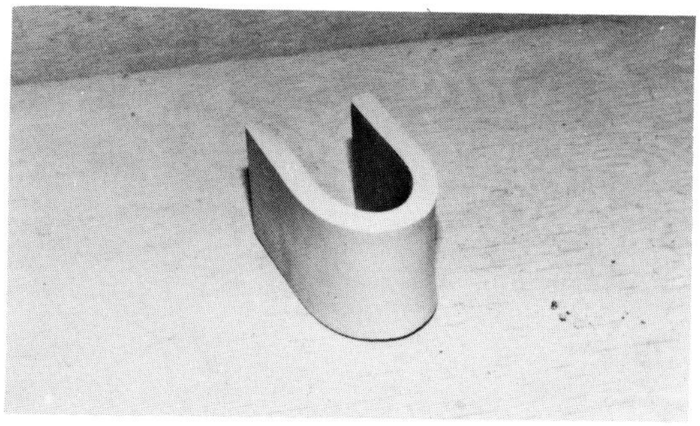

Fig. 57

sawed with the grain, 1/2 inch wide, no welding was done. They were bent double and then pressed flat. Neither of them cracked. That is ductile.

Fig. 58

The specimen in Fig. 59 shows 2 Grade #2 bolts. The one on the left was welded and it cracked. The one on the right was not welded and it did not crack.

Fig. 59

Fig. 60

The bolts in Fig. 60 are ungraded (the square head type). They were both welded. One was a 3/8", the other a 1/2" diameter. They did not crack.

Fig. 61

The 1 inch square tubing in Fig. 61 shows that if you weld across the tubing, it will crack. If you don't weld it crossways, it doesn't crack.

Fig. 62

Fig. 62 shows a crack adjoining a crossweld on a stress member. See a better way on page 26- Fig. 50.

Fig. 63 shows that if you weld around a bolt that comes through a hole, it may break off. Better grades of bolts have markings on top of the head. Grades above one are more likely to break. If you weld the head on the backside, it probably will not affect the bend.

Fig. 63

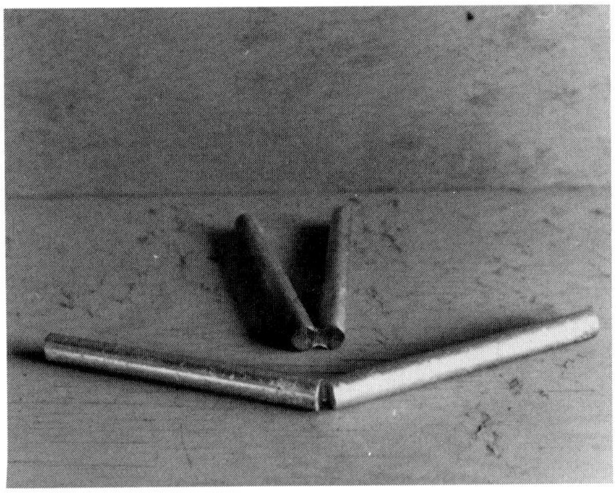

Fig. 64

The above specimens in Fig. 64 were cut from the same 1/2" bar. A stop on the saw was used to cut them the same depth.

The one on the bottom was bent cold to a 38 degree. It cracked because the only place that the metal could stretch was the width of the saw cut at the bottom of the cut.

The specimen on top was bent *cold* also. It was bent cold less than 38 degrees. It had not cracked. Then it was heated red hot and left to cool. When it was cooled to room temperature, it was slightly bent a little more. This procedure was continued until the specimen was bent on a 160 degree. It did not crack. When steel is bent cold, the crystals on the outside stretch. When it is heated to a red heat (non-magnetic), the crystals return to their original shape.

Fig. 65

A very small spot was welded on each of the 3/8" x about 1/2" specimens. The weld area did not get red hot all the way through. The one on the right was heated red hot in the weld area. The one on the left was not. See Fig. 65.

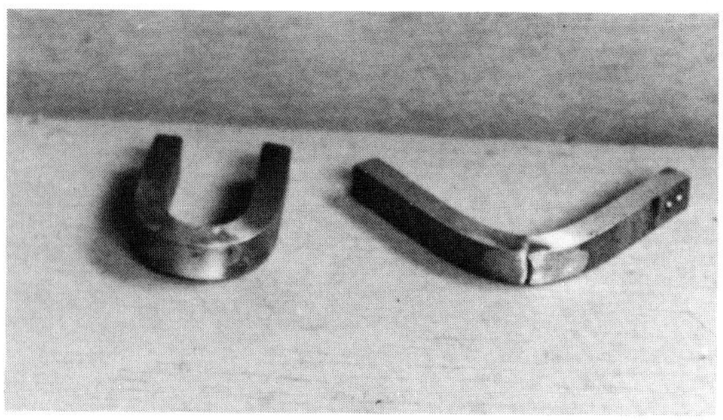

Fig. 66

They were ground and bent. The one that was reheated did *not* crack. The one that was *not reheated* cracked. See Fig. 66.

"Direction of Travel" and "Back Stepping"

Fig. 67

Fig. 67 shows a piece of scrap steel 1/4"x 4 5/8"x 23" long was tacked (welded) on one end to a steel table. A stop was tacked against one side on the other end to show how much the "Direction of Travel" while welding, effects the other end.

One continuous bead was welded left to right across the end that was welded to the table, 1 inch from the welds that held it to the table. The stop was 19 5/16" from the cross weld. When the weld cooled, there was a space .100" or 1/10 of an inch wide between the stop and the 1/4" plate. The 1/4" plate was not cut in two. The welds were made on a solid 1/4" plate.

The stop was knocked off and again placed against the side of the 1/4" plate and tacked.

To show the results of "Back-Stepping", 4 short beads were "back-stepped" across the plate. That is, start about 1 5/32" from the right hand edge of the plate. Travel left to right to the right hand edge. Then start at about the center of the plate. Travel left to right to the first bead, fusing them together. Then another short bead running into the second bead. Then another short bead starting at the left hand edge running into the third bead. (4 back steps).

The bead was 18 1/4" from the stop. When it all cooled, there was only .012" between the stop and the plate.

Then we ran another bead, starting in the center of the plate, traveling left to right to the right hand edge. Then another bead starting at the left hand edge, traveling left to right to the center, meeting the first bead. (2 back-steps) This bead was 17 1/2" from the stop. It moved away from the stop .021". This is only 9 thousandths of an inch more than the 4 back-steps.

On the next bead, we traveled in both directions, starting at the right hand edge traveling right to left, stopping a little short of the center. (The first bead usually pulls more than the second bead.) Then starting at the left hand edge, traveling left to right, meeting the first bead in the center. This bead was 19 1/2" from the stop. It moved away from the stop .002" (2 thousandths of an inch).

I would not recommend starting in the center and traveling to the outside edges. The center would be under compressive stress, and the edges would be under tensional stress. I think it would be more likely that a crack would start on an edge that was under tension stress.

SUMMARY

	Moved	Distance from Stop
Continuous bead L to R	.100"	19 5/16"
Back-stepped 4 times	.012"	18 1/4"
Back-stepped 2 times	.021"	17 1/2"
2 beads R to L to center and L to R to center	.002"	19 1/2"

Why a Weld Bead Shrinks More Where You Stop Welding Than It Does Where You Start

The weld bead and surrounding metal behind the weld crater are hot. The metal in front of the weld crater is cold. This creates a bending force on the workpiece. The cold metal in front of the weld bead is under a compressive stress, (pushing together). As the weld crater moves forward and melts the compressed metal, the weld crater and the red hot metal around it are relaxed. The crystals are rearranged which makes the weld thicker at that point. This is "confined expansion". As soon as the crater solidifies, it starts shrinking which puts **more** compression on the cold metal in front.

Confined Expansion

A condition which resists the free expansion of a weld bead while welding. Now think back to the time that you made your test welds, were the ends of the specimen free to move in and out? Did you weld from edge to edge on the specimen? Did the speicmen get red hot all the way through? If the answers are "yes" that is "free expansion". The specimen can be bent without cracking.

If you weld a crack on an I beam, angle iron, channel iron, etc. the cold metal partly surrounding the weld will resist the expansion. This is "confined expansion".

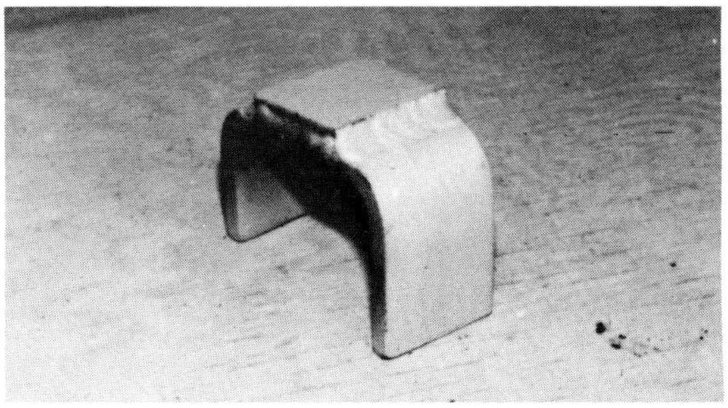

Fig. 68

The specimen in Fig. 68 shows that if it is not too thick, and welded from edge to edge, it can be bent without cracking. It is free to expand in all directions.

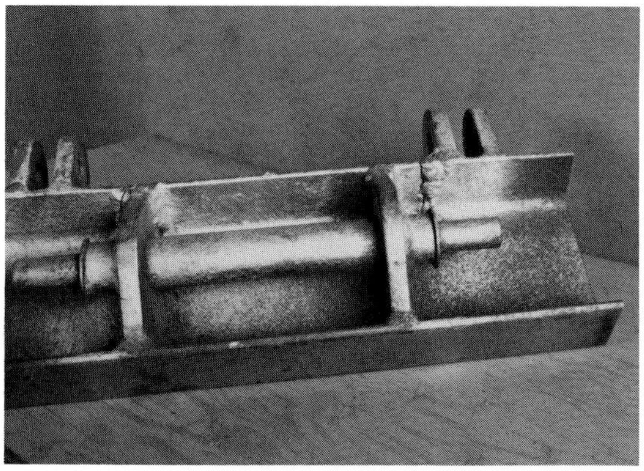

Fig. 69

The specimen in Fig. 69 shows 2 cracks on rewelds, that were welded crossways on a stress member.

Fig. 70

Fig. 70 shows the same thing. If a crossweld is made on a 1 inch square tubing, it will crack.

The Exception to the Rule

Fig. 71

The test weld in Fig. 71 above was made to show the exception to the rule. This test weld was made with no restraints. The ends were free to move outwardly as it was heated, and free to move inwardly as it cooled. The welds are not under as much tensional stress as they would have been if the ends had been welded to a heavy piece or if the patch had been welded in the center of a much larger plate. This piece can be bent without causing a crack adjoining the weld. However, I can't say what would happen under a work-load.

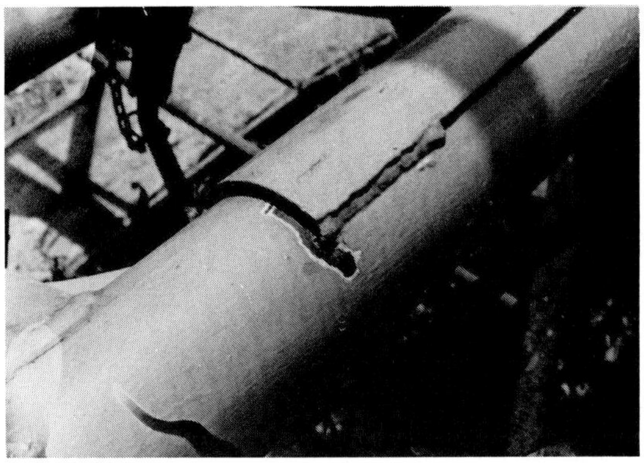

Fig. 72

The above photograph shows a crack (with magnetic particles, see Fig. 72) at the end of a 3 inch weld. The reinforcement is cut on a 90 degree angle. The reinforcement is too wide. It should have been tapered on both sides to about 1 1/2 or 2 inches wide. The 1 1/2 inch width should have been about 3 or 4 inches long and then cut on a 45 degree angle. When a reinforcement wraps half way around a piece of pipe, it makes it rigid. It should not stop abruptly. It causes a shoulder effect which can cause a crack. The shoulder effect and the shrinkage of the weld caused this crack. Both sides of the reinforcement should have been "skip-welded". This means to weld one inch and skip one inch, weld one inch and skip one inch. Repeat all the way on both sides, not the ends.

Magnaflux Testing For Cracks

Magnaflux is a brand name for an Electro magnetic device that assists in finding a crack that cannot be seen with the naked eye. It works like this — place the magnetic poles on each side of the suspected crack. Turn the electric on and spray powdered iron over the surface. The iron dust will stick to the crack. Gently blow the excess away. The magnetic particles will show the crack. **CAUTION:** Do not breathe the iron particles! If you don't have the magnetic device, a large horseshoe magnet will do. If you don't have iron powder, the sawdust from under a metal (steel) saw will do. You may have to sift it to get the smaller pieces. This method also works on a weld that does not go all the way through the work. That is, if you weld over the top of a crack and leave part of the crack below the weld, the dust will stick to the weld bead.

Another way to test for a crack is to clean the surface, pour kerosene over the surface, wipe dry with a clean rag, and then cover with blackboard chalk. The crack will fill with kerosene and when the chalk is applied, the kerosene will wet the chalk to show the crack. You can also buy a kit that works the same way. The kit contains a red liquid dye, and the chalk is brushed on in liquid form.

Water Tank

I remember, back in about 1948, a friend of mine insisted that I help him straighten out a tank that had been screwed up by two beginners. The tank was about 18 or 20 feet in diameter and about 10 or 12 feet high. The 1/4 inch plates had already been rolled to the correct radius. (As you know, the ends of a rolled plate are straight for the distance between the centers on the rolls. They should have been sheared off.) But the beginners had already welded together the bottom of the tank and the first layer of the vertical plates. When they started with the second layer of the plates, (the seams were staggered) the flat places on the ends of the bottom plates took a short cut across the radius of the top plate and left a wide gap horizontally between the top and bottom plates. With a sledge hammer, we drove the seams out to meet the radius on the top plate.

The beginners had already welded some of the second layer plates solidly, except the gap. I remember one place where they had tacked a plate in place and then welded the end vertically from bottom to top, to the end of the other plate. The vertical weld pulled together so much at the top that it broke the tack on the other end and lifted the plate up on the other end, making a gap too large to be welded. (See "Direction Of Travel", pg. 33, fig. 67.) I started welding horizontally from the bottom of their vertical weld toward the loose end. I would weld a couple of inches, then wait for it to cool. (The plate was 10 or 12 feet long.) Weld a couple more inches and wait for it to cool. This method was continued until the short welds pulled the other end back down to the proper place. The end was tacked again and the rest of the weld was "back-stepped" (See pg. 33, fig. 67.) to reduce some of the stresses.

If you have a long bead to run, it is best to back-step it. Back-stepping will not relieve all of the stress, but it will help.

Back-Stepping

Start a bead 6" from the edge of a plate. Travel to the edge of the plate. Start another bead 12" from the edge and weld toward your first bead and into it. Then back up 6" from where you started the last time, each time.

Broken Trailer Frame

I had a truck driver who had a trailer frame welded in Canfield, Ohio. He drove it to Knoxville, Tennessee, and it was broken when he arrived there. When he arrived in Chattanooga, it was broken again. Both welders had welded the ends of the reinforcements. Both breaks were adjoining the end welds. The reinforcements were removed and replaced with a longer one. The ends were *not* welded. It did *not* break again.

Most people agree that: As it cools, every weld shrinks from all directions. When a weld shrinks, the adjoining metal is forced to stretch or crack. If the metal is stretched, it is thinner. Is thin metal as strong as thicker metal?

Some welding instructors teach the beginners to weld crossways on a stress member. They bend a test weld and it doesn't break. This is explained in "The Exception To The Rule" on page 35, fig. 71.

Air Lifts

Years ago, I had some air-lifts installed on my car. They were rubber bags that fit on the inside of the coil springs on the rear of the car. The rear end could be overloaded and then add air pressure to the bags to raise the car up. They worked really well. There was a rubber hose from the bags to the trunk of the car. The air pressure could be adjusted without going under the car. Then one day, one of the bags went flat. It was taken off and cut open to see what caused the crack. There was a reinforcement molded to fit the inside and probably vulcanized to it. The reinforcement was about 1/8 of an inch thick and where it stopped on the sidewall, it made a sharp shoulder. That is where it cracked. If a shoulder will cause a crack on rubber, what will it do to steel?

Subjects of Non-Welding

The following subjects do not pertain to welding. However, a good welder is expected to perform many tasks other than welding. Welding is not a trade, but a sideline of many trades, such as: machinist, iron worker, carpenter, electrician, pipe fitter, etc.

We hope some of you will benefit from this effort. At least, some of them are interesting.

Cover A Padded Bar With Upholstery

This method is for problem applications, where you have a thousand wrestlers a day trying to twist them off. Use a **sponge** rubber pad that is tough (foam will disintegrate). The holes in foam rubber are connected. You can blow smoke through it, but not in sponge.

Glue the pad on to the bar solidly. Leave a 1/4'' gap where the sides of pad come together. Try to place the gap out of sight. The gap will contain the loops and wires, as you will see later.

The padding should be wrapped with strips of cloth to hold it in place until the glue dries. When the glue dries, remove the cloth and determine how wide the cover should be to fit around the padding correctly. Cut a strip of your material 1 inch wide and 2 inches or more longer than the circumference around the padding. Fold 1 inch back and staple it with a paper stapler. Wrap it around the padding with the other end folded back loosely. Bring them together at the gap in the padding. Slip the loose end until the loops or folds touch. This measurement is important. If you like, you may try a piece of scrap material first.

With the material cut to exact size, and sewed lapped over 1 inch on the sides and 1/2 inch on the ends, as shown below fold together, place one end of both sides in the vice and cut both pieces 1/4 inch deep every 1/2 inch, as shown above.

Fig. 73

Apply glue to padded bar, all except 1/2 inch on each side of the gap. If you get glue on the loops, you may not be able to push the wires through. In applying the glue to the cover, raise the flap on the seam up and brush the glue all the way back to the threads, then when your threads rot out, the glue will still be holding.

In placing the cover on the padded bar, line the loops up so that the wire which will be on the inside of the loops will be centered over the gap in the padding.

With the cover glued on, take a piece of 3/32 acetylene welding rod and file one end smooth and bend it 1/2 inch from the end about 1/8 inch off center. Then bend about 2 inches on the other end on a 90 degree for a handle. Push the welding rod through the first loop on the left hand side, then through the second loop on the right hand side, third loop on left, etc.

When you push the wire through one loop, turn the wire about a half turn so that the bent end points to the next loop on the other side. It will start through the loop much easier, and it will pull the cover tight. Push the wire all the way through and out the other end. Thcn push another welding rod through the remaining loops the same way.

On the ends, punch a hole in the cover so you can run a wire around the ends on the inside of the hem. Now go back to the wires that go through the loops. Cut the wires off and file the sharp edges off and with a pair of needle nose pliers, bend the wires and tuck them under the padding. Then twist the wires together that go around the ends, cut them off, file the sharp edges off and tuck them under the padding. Don't twist them too tight. Leave room to tuck them under.

It is important that you cut the material lengthways with the roll instead of crossways — it will stretch more crossways.

When you cut the material, cut it 1 inch longer and 2 inches wider to allow for the hems. Make a straight line 2 inches in from the sides and 1 inch in from the ends. All lines and cuts should be straight. This gives the one who does the sewing something to lap the edge of the material to, which makes the width correct.

Hooks Or Eyes On Spring

To make the hooks or eyes on an expansion spring, place spring in vise vertically, with one round of the wire above the jaws. Use a pair of duck bill pliers to twist the top round straight up. You may have to hold the second round down with a screwdriver. Now the hook is standing straight up and you have not bent the wire. You have twisted it a quarter turn as shown on one end below and your hook is on one side of the

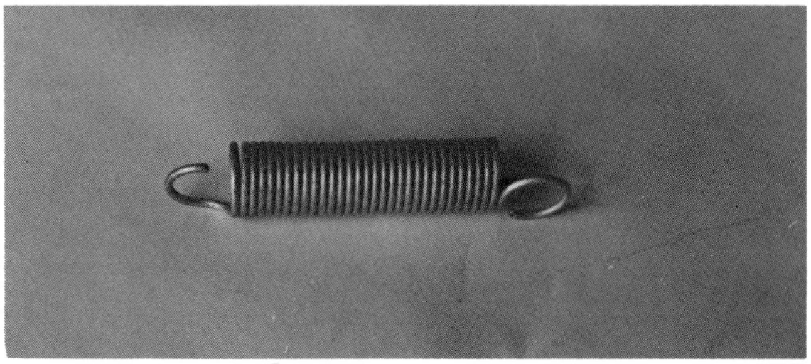

Fig. 74

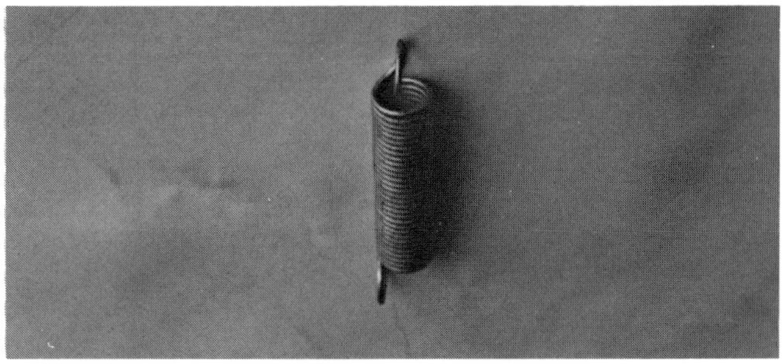

Fig. 75

spring. To center the hook, twist it one quarter turn to the center of the spring as shown above. This method does not bend the spring steel wire, it only twists, it. You can also use this method to replace a broken off hook or eye.

Make A Coil Spring

In an emergency, you can make a coil spring. You will need some spring steel wire of the correct size. You will need a mandrel the right size.

Drill a hole slightly larger than your wire, 1/4 inch from the end. Then saw a slot with a hacksaw, slightly larger than your wire, a few degrees clockwise of your hole as shown below.

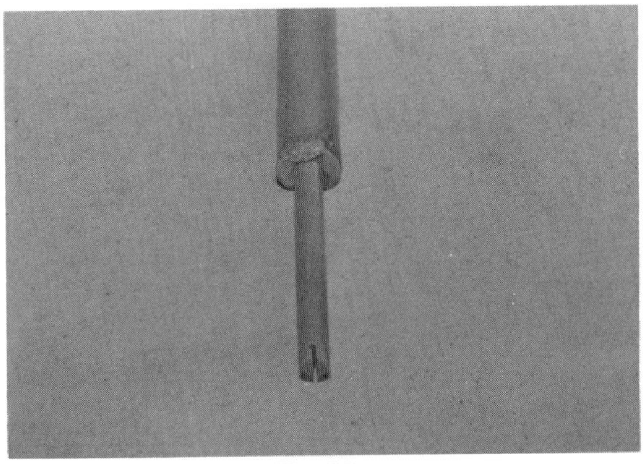

Fig. 76

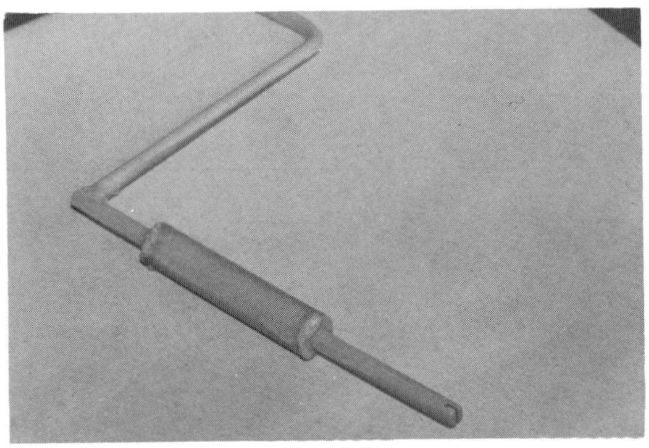

Fig. 77

If you don't have a lathe, cut a piece of pipe 2 inches longer than the jaws of your vise. Weld a washer on each end of the pipe with holes drilled in the washers to fit the mandrel. Weld a handle on the other end of your mandrel to turn it with as shown above. You will need someone else to turn the crank with the pipe held in the vise. Place the end of your spring wire in the slot and bend it. With the mandrel turning, pull tightly on the wire. For an expansion spring, wind the wire close together. For a compression spring, lead the wire away from the last round so the spring will have room to compress. When you finish, turn the crank the other way to relieve the tension. The spring will be much larger in diameter than the mandrel.

Cooling Fin

I was once told by an engineer that a 7'' aluminum cooling fin placed on a 4'' steel hydraulic cylinder, would contract when heated, which would make it tighter on the cylinder. I didn't believe it. Aluminum expands more than steel. Then I asked another engineer. He agreed with the first one. He explained that it would expand both ways from a point slightly to the inside of the center between the two circumferences, and it would tighten on the cylinder. I still didn't believe it, so I pressed one fin onto a cylinder. With the cylinder standing vertically, heat was added to the inside of the cylinder. When the fin got hot, it dropped. I think the aluminum expanded in **all** directions, causing the hole to be larger as well as the outside circumference. After all, the sides of the fin are only 1 1/2 inches and the circumference is over 21 inches.

Shaft Will Break At Shoulder

Figure 78 shows a broken shaft at the end of a rod end. It broke for two reasons: partly because the threads are the weakest place and partly because the rod end was not tapered enough to give. It broke off right at the end of the lock nut. That is a notch effect.

The same thing happens when you saw a half-inch rod with a hacksaw, leaving only 1/8 inch holding. Bend it and it breaks off. That is because the metal can only stretch in one place, the width of your saw blade. Take a 1/8 inch welding rod and try to break it. It bends in a much larger radius.

Years ago electric irons came with a drop cord the same size all the way. With all the flex required for ironing clothes, the cords inevitably broke off right at the plug end that attaches to the iron. A heavy rubber sheath, tapered out six or eight inches caused the cord to bend in a much larger radius and the breaking stopped.

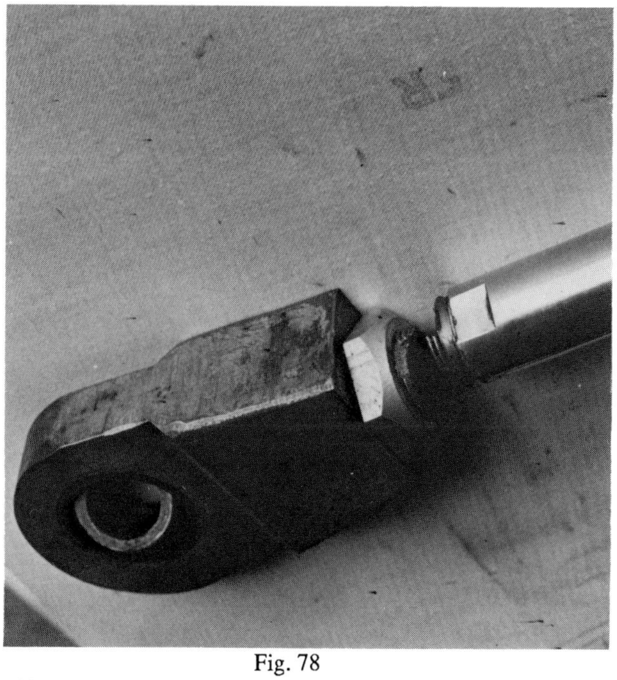

Fig. 78

The same principle applies in both cases.

In 20 years, I had 3 rods to break as shown in Fig. 78 (Shaft will break at shoulder).

Then I made new ones, 1/8 of an inch in diameter larger. To avoid cutting as deep into the shaft, I used 18 threads per inch instead of 12. At the end of the threads, the shaft was turned down about 1/2 inch in length to the size of the bottom of the threads to let it bend in a larger radius. Then it was filed, sanded and polished to remove any cross scratches that might start to crack. They have been in service about 6 years with no failures.

How To Temper Small Parts

Using water hardening drill rod, cut to desired lengths. Heat them to a red heat and plunge them into water. This makes them hard and brittle. Take them out of the water and sand the oxide off so you may be able to see the heat colors. Heat a piece of heavier steel to a red heat, and place the parts on top of it. Watch the heat colors. If you quench them when they first turn dark blue, they will be hard enough to break. If you want them to be soft enough to bend, wait a few seconds until they turn light blue then quench them.

Rod Ends

If you want to pin a threaded rod end to a rod to prevent it from turning, and the rod-end and pin still be interchangeable with the other rods, here's how. Weld a piece of steel onto a nut to hold a hardened drill bushing. The hole for the pin should be near the end of the rod. Screw the nut onto the rod until it is flush with the end of the rod. Then place the rod-end against the rod and start turning the rod-end, nut and drill bushing all together. There may be a space between the nut and the rod-end. If the space is wider than the distance between two threads, take it off. You held the rod-end crooked and didn't catch the first thread. Screw the rod-end on to the desired depth and drill hole through drill bushing into and out the other side of the rod-end and rod.

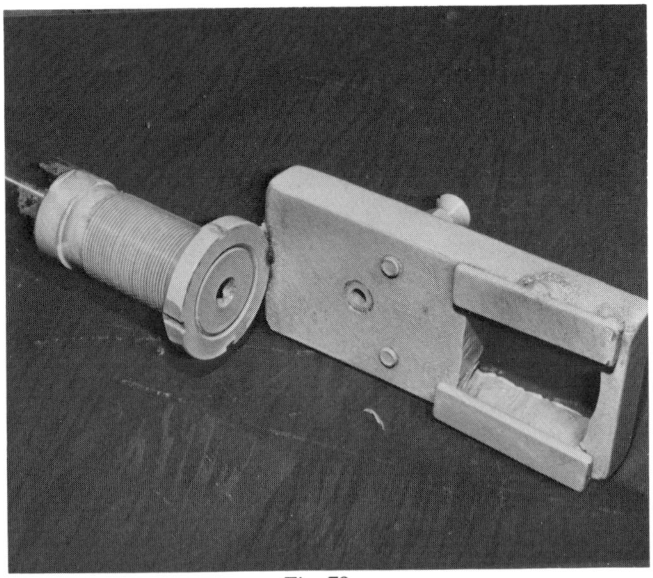

Fig. 79

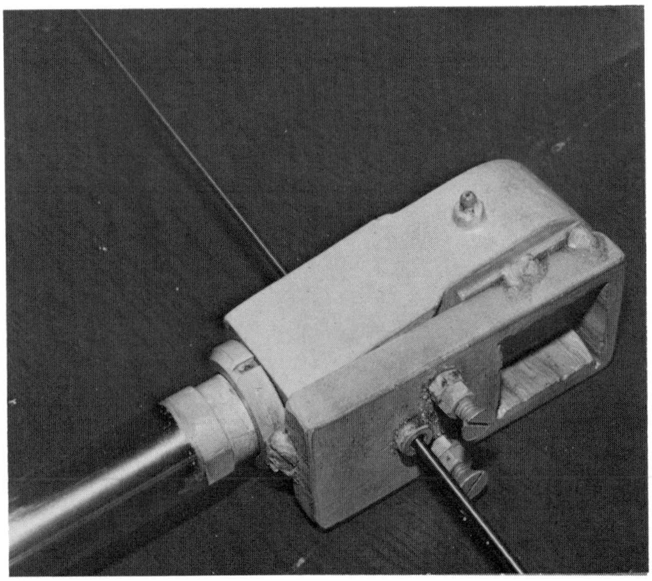

Fig. 80

It is important to have a lock-nut **tightly** against the rod end. I had a nut get loose and wore the threads on the rod-end too much. I shrank the rod-end at the threads (as explained in "How To Shrink A Steel Pulley") a little too much. I had to tap it out to get it on.

Find Diameter of Numbered Machine Screws

Multiply number of screw by .013 then add .060.

Example: The screws that hold the cover plates on your electric wall switches at home are 6-32.

$$\begin{array}{r} .013 \\ \times 6 \\ \hline .078 \\ +.060 \\ \hline .138 \end{array}$$

Diameter of screw is 138 thousandths of an inch

Which Way To Turn A Turnbuckle

Look at the two threads on this turnbuckle. If the threads on one of them look like they are slanted more than the other ones, they are the left-handed threads. Facing either end of the bolt, turn a left-handed nut counter clockwise to move it away from you. Turn a right-handed nut clockwise to move it away from you.

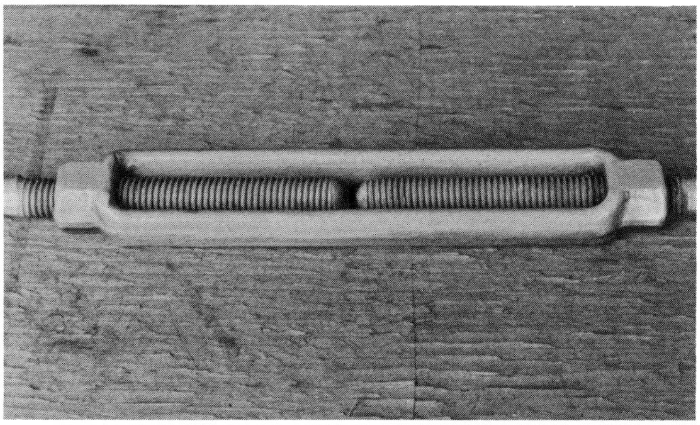

Fig. 81

Dish In Wheel

Almost every wheel has a dish in it. If the hub was not off-center to the rim, it would be like a square tank to hold pressure. A flat surface bends more than a curved surface when pressure is applied. A welder once asked me if I could build a square hot water tank. I said, "yes, but it would take a lot of stay-bolts to prevent the sides from bulging." He said he built one and the next morning it was almost round. Same thing goes for a retaining wall. A straight wall sometimes falls over or breaks. A curved or zig-zag wall is much stronger.

Try to stand a flat card on edge. Then bend it and see the difference.

To make a flat frame rigid, cross-brace from corner to corner. Where the braces cross in the center, they must be bent to make that point higher than the frame. The cross must be bolted or welded to prevent sliding.

Blowing Grip On Bar

The picture below shows a steel bar 3/4" in diameter. The plastic grip has a 5/8" hole in it. The grip was forced on with air pressure. Notice that the grip is not on all the way. The end of the air nozzle hit the end of the bar and stopped. This method may be used to cover a pipe with rubber or plastic tubing. The end of the pipe must be sealed and one end of the rubber tubing fastened to the air nozzle air tight. The rubber tubing will slide over the pipe on a layer of escaping air.

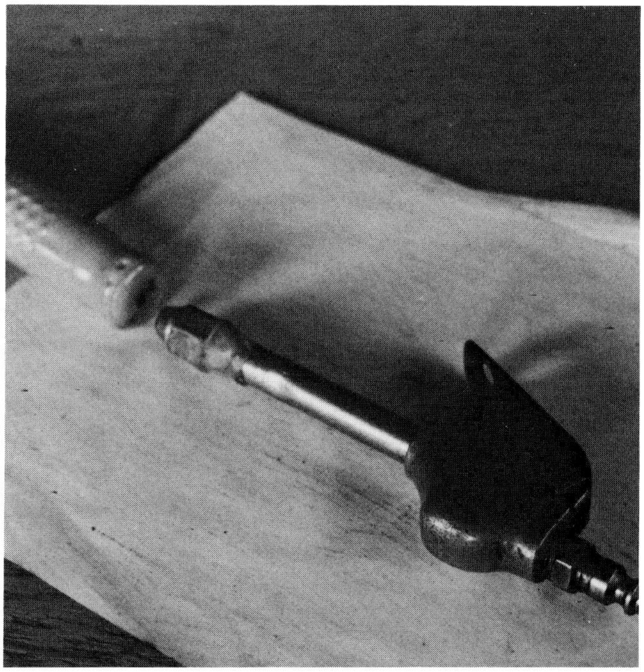

Fig. 82

How To Find A Tap Drill Size Without A Chart

Example: Find a drill size for 1/2 inch 13 thread tap. Divide number of threads per inch into one inch.

```
        .0769
 13 ⟌ 1.000
        91
         90
         78
         120
         117
```

Then subtract .0769 from tap size — $\frac{1}{2''}$ = .500

```
.500
.0769
.4231 = drill size
```

$\frac{1}{64''}$ = .0156 — then divide .0156 into .4231

```
              27
.0156 ⟌ .4231       27             .4231     27
          312       64                64     64''
          1111                     16924
          1092      OR             25386
            19                   27.0784
```

In other words: divide the number of threads per inch into one inch and subtract the answer from the tap size. Then find the nearest standard drill size. A few thousandths does not matter.

Finding Half Of A Mixed Common Fraction

Sometimes it is necessary to find the center of something. That is — find half of the total length.

Example: Whole number **ODD**

$$5\frac{1}{2} \div 2 = 2\frac{3}{4}$$

DISCARD the odd whole number making the 5 a 4. Divide 4 by 2 and place here.

Multiply the denominator by 2 and place here.

Add the numerator and the denominator together and place here.

Example: Whole number **EVEN**

Divide the whole number by 2 and place here

$$4\frac{1}{2} \div 2 = 2\frac{1}{4}$$

Multiply denominator by 2 and place here.

MOVE numerator to here.

In other words: Take the odd whole number and throw it away — forget about it. That changes the 5 into a 4. Take half of the 4 and write it down, 2 - with a dash to the right of it. Then add the top number and the bottom number of the fraction together and place it above the dash $2\,\underline{3}$, then double the bottom number of the fraction and place it below the dash this way: $2\frac{3}{4}$.

$$37\frac{21}{32} \div 2 = 18\frac{53}{64}$$

If someone tells you something that you don't already know, don't even think about making a defensive statement. You are not supposed to know everything. Nobody else does. Check it out and see Who Is Right.

However, you can't believe everthing you see, hear, or read.

Facts

Nickel is magnetic.

U.S. Paper Currency is slightly magnetic in places. When you turn the steering wheel on your car to the left, the left front wheel turns more than the right one does, and vice versa.

The file and scratch test is a quick procedure for checking hardness. Shopmen frequently use this method to predict the machinability of a metal. It consists of simply trying to scratch or cut the surface of a metal with a file or pointed object of known hardness. All scratch tests reveal only a superficial or outer skin hardness. They tell us nothing about the hardness 1/8 in. below the surface.

Table 3-2 gives a relation between Brinell hardness and the hardness as estimated with a machinist's new hand file.

Table 3-2. File Hardness Test Data

BRINELL HARDNESS	FILE ACTION
100	File bites into surface very easily
200	File removes metal with slightly more pressure
300	Metal exhibits its first real resistance to the file
400	File removes metal with difficulty
500	File just barely removes metal
600	File slides over surface without removing metal. File teeth are dulled.

Simple Hardness Tester

Welders sometimes make their own set of hardness testers, by heat treating 1/4-in. high carbon steel electrodes to give a graduated series of hardnesses.The rods are labeled with their respective Rockwell C hardnesses. A sharp point is then ground on the end of each rod, taking care not to soften the steel by overheating.

By drawing these scratch testers, one at a time, over a smooth metal surface, the welder can determine which rod has a hardness about the same as the metal being scratched. If the sharp point of the scratch tester digs in and leaves a scratch, its hardness is greater than that of the metal being tested. If the point only slides over the surface without scratching, the scratch tester is softer than the metal tested.

The hardness set is inexpensive and is convenient to use in hard-to-reach places, where a conventional test would be impossible.

Fusibility: The Ease of Melting

Fusibility is a measure of the ease of melting. Mercury (the metal with the lowest melting point) melts at -38 degrees Fahrenheit, while tungsten, which has the highest melting point, melts at 6,100 F.

A pure metal has a definite melting point, which is the same temperature as its freezing point. Alloys and mixtures of metals, however, have a temperature at which melting starts and a higher temperature at which the melting is complete.

Figure 3-9 gives the melting point of a few metals and other temperatures of interest.

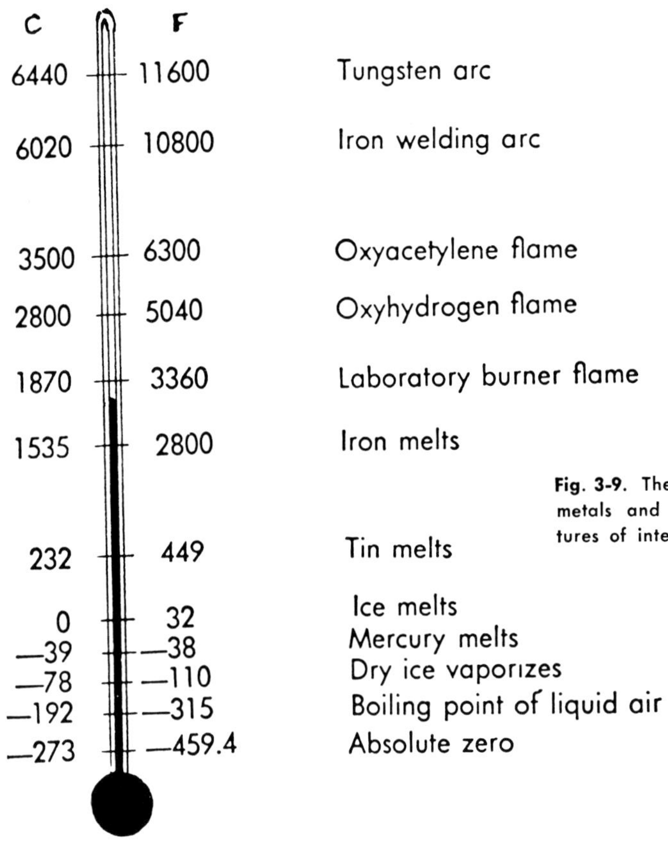

Fig. 3-9. The melting points of some metals and a few other temperatures of interest for comparison.

Reprinted With Permission From: The James F. Lincoln Arc Welding Foundation

Heat of Fusion

The *heat of fusion* is the quantity of heat necessary to change one pound of a solid material to a liquid without temperature change.

The British thermal unit (Btu) is used to measure the quantity of heat; for all practical purposes, it is the amount of heat required to raise the temperature of 1 pound of water 1 degree Fahrenheit.

The heat of fusion of ice is 144 Btu per lb. In comparison, here are the heats of fusion of a few metals:

Table 3-3. Heat of Fusion of Metals

Aluminum	170 Btu per lb	Tungsten	79 Btu per lb
Magnesium	160 " " "	Silver	45 " " "
Chromium	136 " " "	Zinc	43 " " "
Nickel	133 " " "	Gold	29 " " "
Molybdenum	126 " " "	Tin	26 " " "
Iron	117 " " "	Lead	11 " " "
Manganese	115 " " "	Mercury	5 " " "
Copper	91 " " "		

More heat is required to melt a pound of ice than is needed to melt a pound of iron (provided of course that both materials are at their melting point). The figures are 144 Btu for water and 117 Btu for iron. It takes more heat, however, to melt a pound of aluminum than a pound of ice—170 Btu as against 144 Btu.

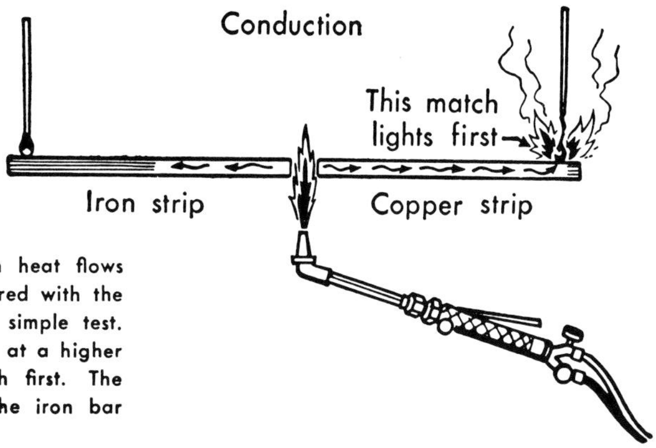

Fig. 3-10. The rate at which heat flows through copper can be compared with the same property in iron by this simple test. The metal which conducts heat at a higher rate (copper) lights the match first. The match at the outer end of the iron bar will burst into flame later.

Thermal Conductivity

Thermal conductivity is a measure of the rate at which heat will flow through a material. The difference in thermal conductivity between iron and copper is easily demonstrated, Fig. 3-10. The copper conducts heat much faster than does the iron.

If one end of a copper bar is kept in boiling water (212 F) and the other end in chipped ice, heat will flow into the bar from the water, then through the bar and to the ice, causing it to melt. The rate at which the ice melts indicates the rate heat is flowing through the bar.

The amount of ice that melts depends upon:

1. Time	The longer the time, the more ice will melt.
2. Size of the bar	The larger the cross-sectional area of the bar, the more heat will flow.
3. Length of the bar	The shorter the bar, the faster the ice will melt.
4. The temperature to which the bar is heated	The higher the temperature of the hot end of the bar, the faster the ice will melt.
5. Thermal conductivity	The higher the thermal conductivity of the bar, the more heat will flow.

The amount of heat flow, therefore, depends upon time, area, length, temperature difference and thermal conductivity.

Table 3-4. Thermal Conductivity of Metals

Metal	Chemical Symbol	Btu/sq ft/in./hr/°F	Relative Conductivity Based on Silver as 100%
Aluminum	Al	1428	49.7%
Copper	Cu	2664	92.7
Gold	Au	2037	70.9
Iron-Pure	Fe	467	16
Iron-Steel		313	10.9
Iron-Cast		316	11
Lead	Pb	241	8
Mercury	Hg	476	17
Molybdenum	Mo	1004	34.9
Nickel	Ni	412	14.3
Platinum	Pt	483	16.8
Silver	Ag	2873	100
Tin	Sn	450	15.6
Tungsten	W	1381	48
Zinc	Zn	770	27

Thermal Expansion

Thermal expansion is the increase in dimensions of a body due to a change in its temperature. See Fig. 3-11 for thermal expansion of selected materials.

The coefficient of linear expansion is the ratio of the change in length of a material, caused by heating it one degree, divided by the original length.

The coefficient of linear expansion of iron at room temperature is 0.0000065 per degree F. ($6.5 \times 10^{-6}/°F$).

The total increase in length of an iron bar 100 feet long which is heated from 10 F to 110 F will be:

$0.0000065 \times (110\text{-}10) \times 100 = 0.065$ ft or 0.78 in.

The coefficient of cubical expansion equals approximately three times the coefficient of linear expansion.

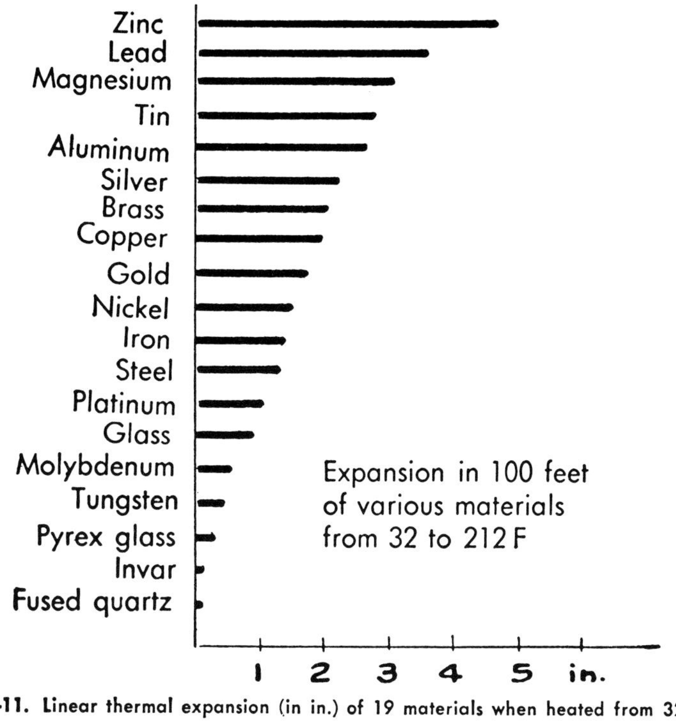

Fig. 3-11. Linear thermal expansion (in in.) of 19 materials when heated from 32 F to 212 F.

Reprinted With Permission From: The James F. Lincoln Arc Welding Foundation

Table 4-1. Typical Ferrous Materials

MATERIAL	SPEC.	CHEMICAL ANALYSIS: CARBON	CHEMICAL ANALYSIS: OTHERS	MECHANICAL PROPERTIES: YIELD STRENGTH	MECHANICAL PROPERTIES: TENSILE STRENGTH	MECHANICAL PROPERTIES: ELONGATION % IN 2 IN.	MECHANICAL PROPERTIES: BRINELL HARDNESS
Cast Iron, Gray, Grade 20	ASTM A48-56	3.00-4.00		. . . . .	20,000	. .	163
Gray Grade 30	ASTM A48-56	3.00-4.00		. . . . .	30,000	. .	180
Nickel		2.00-3.50	Ni 0.25-0.50	. . . . .	40,000	. .	310
Chrome-Nickel		2.00-3.50	Ni 1.00-3.00 Cr 0.50-1.00	. . . . .	53,000	. .	510
White		2.00-4.00	Si 0.80-1.50	. . . . .	46,000	. .	420
Malleable	ASTM A47-52	1.75-2.30	Si 0.85-1.20	35,000	53,000	18	140
Iron, Wrought, Plates	ASTM A42-55	0.08	Si 0.15 Slag 1.20	26,000	46,000	35	105
" " Forgings	ASTM A75-55	0.01-0.05	Fe 99.45-99.80	25,000	44,000	30	100
Steel, Cast, Low Carbon		0.11	Mn 0.60 Si 0.40	35,000	60,000	22	120
" " Medium Carbon		0.25	Mn 0.68 Si 0.32	44,000	72,000	18	140
" " High Carbon		0.50		40,000	80,000	17	182
Steel, Rolled, Carbon	SAE 1010	0.05-0.15		28,000	56,000	35	110
" " Carbon	SAE 1015	0.15-0.25		30,000	60,000	26	120
" " Carbon	SAE 1025	0.20-0.30		33,000	67,000	25	135
" " Carbon	SAE 1035	0.30-0.40		52,000	87,000	24	175
" " Carbon	SAE 1045	0.40-0.50		58,000	97,000	22	200
" " Carbon	SAE 1050	0.45-0.55		60,000	102,000	20	207
" " Carbon	SAE 1095	0.90-1.05		100,000	150,000	15	300
" " Nickel	SAE 2315	0.10-0.20	Ni 3.25-3.75	90,000	125,000	21	230
" " Ni.-Cr.	SAE 3240	0.35-0.45	Ni 1.50-2.00 Cr 0.90-1.25	113,000	136,000	21	280
" " Moly.	SAE 4130	0.25-0.35	Cr 0.50-0.80 Mo 0.15-0.25	115,000	139,000	18	280
" " Cr.	SAE 5140	0.35-0.45	Cr 0.80-1.10	128,000	150,000	19	300
" " Cr.-Va.	SAE 6130	0.25-0.35	Cr 0.80-1.10 Va 0.15-0.18	125,000	150,000	18	310
" " Si.-Mn.	SAE 9260	0.55-0.65	Mn 0.60-0.90 Si 1.80-2.20	180,000	200,000	12	390

Reprinted With Permission From: The James F. Lincoln Arc Welding Foundation

Steels in the low-carbon group are generally tough, ductile and easily formed, machined and welded. Although low-carbon steels are not especially hard, some grades respond to heat treatment and are readily case-hardened by carburizing, cyaniding, flame-hardening, etc.

Table 4-2. Uses for Steel by Carbon Content

CARBON CLASS	CARBON RANGE %	TYPICAL USES
Low	0.05 - 0.15	Chain, nails, pipe, rivets, screws, sheets for pressing and stamping, wire.
	0.15 - 0.30	Bars, plates, structural shapes.
Medium	0.30 - 0.45	Axles, connecting rods, shafting.
High	0.45 - 0.60	Crankshafts, scraper blades.
	0.60 - 0.75	Automobile springs, anvils, bandsaws, drop hammer dies.
Very High	0.75 - 0.90	Chisels, punches, sand tools.
	0.90 - 1.00	Knives, shear blades, springs.
	1.00 - 1.10	Milling cutters, dies, taps.
	1.10 - 1.20	Lathe tools, woodworking tools.
	1.20 - 1.30	Files, reamers.
	1.30 - 1.40	Dies for wire drawing.
	1.40 - 1.50	Metal cutting saws.

Medium-carbon steels—have a carbon range of 0.30 to 0.45%. They are strong, hard and not so easily forged or welded as low-carbon steels. If extensive welding is to be done on medium-carbon steels, it is desirable to use those steels at the lower end of the carbon range—between 0.30 and 0.35% carbon. As the carbon content increases above 0.35%, the steel becomes increasingly difficult to weld since there is a greater tendency toward brittleness of the weld. Special electrodes and procedures are often necessary to prevent weld cracking.

High-carbon (0.45 to 0.75%) and **very-high-carbon** (0.75 to 1.5%) **steels**—are very strong and hard. Both properties increase with an increased carbon content.

While carbon has an important influence on the characteristics of steel, the degree to which impurities are not removed by refining is also important. A slight increase in the percentage of phosphorus or sulphur will materially lower the ductility, malleability, fatigue and shock resistance and welding qualities of a steel.

High- and very-high-carbon steels respond well to heat treating. Nearly any degree of hardness, temper or strength may be obtained. In the annealed state, most of these materials may be readily machined. They also may be hot-worked for forming.

Fig. 4-1. A great many different shapes of rolled carbon and low-alloy steels are available from steel warehouses.

Reprinted With Permission From: The James F. Lincoln Arc Welding Foundation

Table 6-1. AWS Electrode Classification System

DIGIT	SIGNIFICANCE	EXAMPLE
1st two or 1st three	Min. tensile strength (stress relieved)	E-60xx = 60,000 psi (min) E-110xx = 110,000 psi (min)
2nd last	Welding position	E-xx1x = all positions E-xx2x = horizontal and flat E-xx3x = flat
Last	Power supply, type of slag, type of arc, amount of penetration, presence of iron powder in coating	See Table 6-2

Note: Prefix "E" (to left of a 4 or 5-digit number) signifies arc welding electrode

Measuring Preheat and Interpass Temperatures

Instruments are available for measuring surface temperatures. One such instrument has a wire coil, which is placed on the work; within a few seconds the temperature may be read on the dial at the handle end of the pyrometer. Preheat temperatures, however, can be satisfactorily estimated without expensive instruments.

Several suppliers* offer materials that melt over comparatively narrow ranges of temperatures. These indicators are furnished in sticks, liquids or pellets. If a preheat temperature of, say, 400 F, is wanted, the 400 F stick is scratched over the surface of the metal; if it leaves a white line, the metal is not at 400 F. If a transparent liquid line is formed, the metal is at or above the temperature printed on the stick.

Some common items used as rough temperature indicators are listed here:

1. Blue chalk: A mark made with carpenter's blue chalk on hot metal will turn whitish gray when the metal is above 625 F.
2. Red lumber-marking crayon: Its mark melts at about 615 F. The red mark turns pink at about 635 F and turns a dirty gray at about 650 F.
3. Solder: 50-50 solder (50% lead and 50% tin) starts melting at 360 F and is completely molten at 420 F.
4. Pine stick: White pine stick chars at about 635 F (used to determine the proper pouring temperature for some babbitts).

Another method for checking preheat temperatures is by temper colors: The temperature and accompanying color that appears on a freshly filed surface of carbon steel are given in Table 13-1.

Table 13-1. Temper Colors

TEMPERATURE, F	COLOR
400	Faint straw
440	Straw
475	Deep straw
520	Bronze
540	Peacock
590	Full blue
640	Light blue

It should be noted that the presence of alloying elements in the steel affects the colors formed, and some of the higher alloy steels have no temper colors.

*The Tempil Corporation; Markal Company; and Curtiss-Wright Corporation.

Appendix 2—MELTING POINTS OF SOME ELEMENTS

ELEMENT	CENT.	FAHR.	ELEMENT	CENT.	FAHR.
Hydrogen	—259	—434	Uranium	1,132	2,070
Oxygen	—218	—360	Manganese	1,260	2,300
Nitrogen	—210	—346	Beryllium	1,277	2,332
Phosphorus	44	111	Silicon	1,420	2,588
Sodium	97	207	Nickel	1,452	2,646
Sulphur	113	236	Cobalt	1,480	2,696
Selenium	218	424	Chromium	1,520	2,768
Tin	232	450	Iron	1,530	2,786
Bismuth	271	520	Titanium	1,668	3,035
Cadmium	321	610	Thorium	1,750	3,182
Lead	327	621	Zirconium	1,852	3,366
Zinc	419	737	Vanadium	1,900	3,450
Antimony	630	1,166	Hafnium	2,222	4,032
Magnesium	651	1,204	Boron	2,316	4,200
Aluminum	659	1,218	Columbium (Niobium)	2,468	4,474
Calcium	810	1,490	Molybdenum	2,610	4,730
Barium	850	1,562	Tantalum	2,996	5,425
Silver	960	1,761	Tungsten	3,410	6,170
Copper	1,083	1,981	Carbon	3,828	6,700

NOTES

NOTES